The Scientific Revolution

THE CONTROL OF NATURE
Series Editors:
Margaret C. Jacob and Spencer R. Weart

PUBLISHED

SCIENTISTS AND THE DEVELOPMENT OF NUCLEAR WEAPONS
From Fission to the Limited Test Ban Treaty, 1939–1963
Lawrence Badash

TECHNOLOGY AND SCIENCE IN THE INDUSTRIALIZING NATIONS, 1500–1914
Eric Dorn Brose

EINSTEIN AND OUR WORLD
David Cassidy

NEWTON AND THE CULTURE OF NEWTONIANISM
Betty Jo Teeter Dobbs and Margaret C. Jacob

THE SCIENTIFIC REVOLUTION
Aspirations and Achievements, 1500–1700
James R. Jacob

TOTALITARIAN SCIENCE AND TECHNOLOGY
Paul R. Josephson

CONTROLLING HUMAN HEREDITY
1865 to the Present
Diane B. Paul

FORTHCOMING

GENDER AND SCIENCE
Paula Findlen and Michael Dietrich

CONTROL OF NATURE

The Scientific Revolution

Aspirations and Achievements, 1500-1700

James R. Jacob

HUMANITIES PRESS
NEW JERSEY

First published in 1998 by Humanities Press International, Inc.
165 First Avenue, Atlantic Highlands, New Jersey 07716

Library of Congress Cataloging-in-Publication Data

Jacob, James R. (James Randall), 1940–
The scientific revolution : aspirations and achievements, 1500–1700 / James R. Jacob.
p. cm. — (The control of nature)
Includes bibliographical references and index.
ISBN 0-391-03977-6. — ISBN 0-391-03978-4 (pbk.)
1. Science, Renaissance. 2. Science—France—History—17th century. 3. Science—England—History—17th century. I. Title. II. Series.
Q125.2.J33 1997
509'.4'0903—dc21 97-17851
CIP

Printed in the United States of America

10 9 8 7 6 5 4 3 2 1

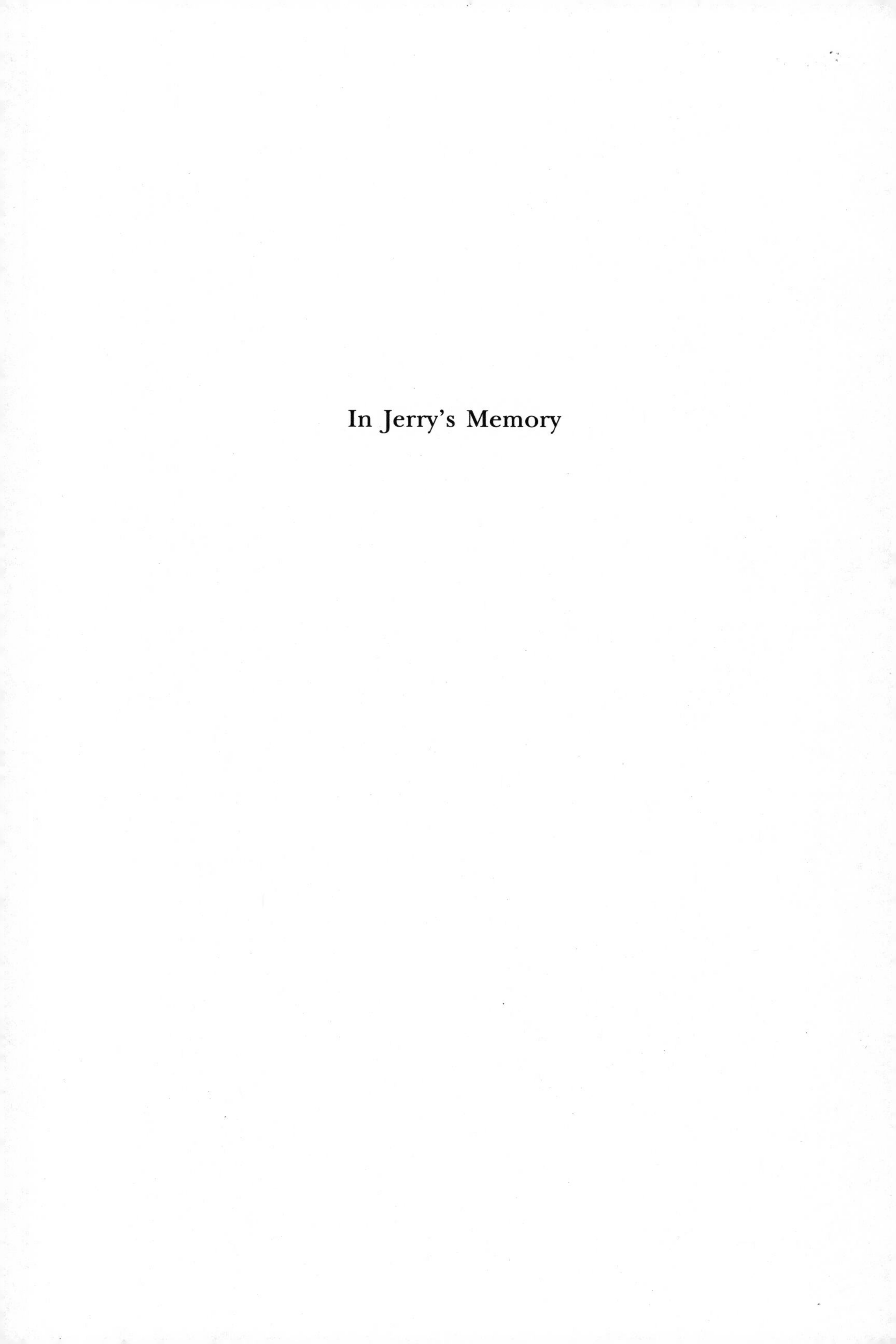

In Jerry's Memory

Contents

List of Illustrations

Series Editors' Preface

THIS SERIES OF historical studies aims to enrich the understanding of the role that science and technology have played in the history of Western civilization and culture and, through that, in the emerging global civilization. Each author has written with students and general readers, not specialists, in mind, and the volumes have been written by scholars distinguished in their particular fields. In this book, James R. Jacob—well known as one of the founders of the social history of science, who has pioneered work on Robert Boyle and also on the origins of the Enlightenment in England—draws upon his expertise to address the central issue in the Western history of science, namely the emergence of the Scientific Revolution from roughly 1540 to 1687, the year when Newton published the *Principia*.

The aim of this book on the Scientific Revolution is not to debunk the idea that it never happened—now fashionable in some quarters—but to lay out the vast change in our approach to nature and truth that occurred in the early modern European era. It ranges back to the ancient writings that offered inspiration, but it also describes how dramatically new understandings of nature took shape largely in the seventeenth century, in both England and on the Continent. The volume also engages with the questions and debates that inform contemporary scholarship. Taking a sample of the many complex processes that scholars are currently investigating, it seeks to make the science accessible and the natural philosophy—as science was then called—readable by anyone.

The current debates and overall perspective in this volume, and the series to which it belongs, emphasize the "Control of Nature." While not excluding a discussion of how knowledge itself develops, how it is assembled through the interplay of research into nature with the values and beliefs of the researcher, this volume looks primarily at how science and technology interact with religious, social, economic, and political life, in ways that transform the relationship between human beings and nature. In every volume in this series we are asking the student to think about how the modern world came to be invented, a world where the call for progress and the need to respect humanity and the environment produce a tension, on the one hand liberating, on the other threatening to overwhelm human resources and ingenuity. The early scientists and philosophers whom you will meet here could not have foreseen the kind of power that modern science and technology both promise and deliver. But they were also dreamers and doers—even shrewd promoters—who changed forever the way people view the natural world.

MARGARET C. JACOB
ROSEMARY STEVENS
SPENCER WEART

Introduction

THIS BOOK IS meant to be an introduction to a large and complicated subject that has come to be called the Scientific Revolution. This refers to the fundamental changes in our understanding of the natural world that occurred in the sixteenth and seventeenth centuries and that led to a rejection of ancient and medieval thinking about the universe in favor of the new thinking that gave birth to modern science. This short book does not pretend to tell the whole story of this momentous transformation, perhaps more important than any other in modern history. But it does highlight and survey what are often considered to be the six principal developments associated with this shift from old to new science. The six changes that will be focused on are

1) the abandonment of an ancient Greek picture of a finite, earth-centered universe and its replacement by the modern picture of a solar system surrounded by an infinite universe;
2) the gradual rejection of the Aristotelian binary physics of metaphysical "natures" and "places" in favor of the modern physics of universal forces whose behavior can be quantified and expressed in terms of mathematical laws;
3) the medical revolution that culminated in the discovery of the circulation of the blood, undermined the authority of ancient Greek medicine, and put animal (and human) physiology on a new foundation;
4) the shift from an Aristotelian theory of knowledge, which was confident of the truth of what we perceive, to a modern skepticism that doubts our capacity to know truth but nonetheless finds the resources, intellectual and cultural, to overcome pessimism and to insist that scientific knowledge is still attainable and that this pursuit, "the advancement of learning," is exceedingly important and worthwhile (Bacon 1955, 157);

5) the development of new methods for establishing scientific certainty by relying on mathematics and on techniques and instrumentation for guarding against and compensating for the fallibility of both reason and the senses; and, finally,
6) the founding of the world's first national, government-sponsored scientific societies for promoting collaborative research, spreading scientific knowledge, and stimulating inquiry.

Because this book is meant to be very brief, drastic decisions had to be made about what to put in and what to leave out. First, I did not want the book to be a withering barrage of names and dates, as would have come from any attempt to make it comprehensive. I have focused on the major figures and developments so that I could still flesh these out to a limited degree and make my account more readable and digestible. This focus also gives the book a thematic coherence that it would not otherwise have. Second, I have intended it for readers without any scientific background and have made it as nontechnical as I thought possible. I have summarized, in plain English, the profound contributions of mathematical physics on which the Scientific Revolution was built. For brevity's sake, I have also drastically curtailed discussion of the history of medicine and all but excluded the history of technology. This book is meant to be a reliable but animated beginner's handbook, and I have written it this way because, in teaching the subject at both graduate and undergraduate levels for many years, I have found such books to be as scarce as they are valuable.

Third, modern science grew out of various, often competing and long-lived intellectual traditions, and so I have shown throughout how science evolved by both rejecting *and* adapting these streams of thought. Hence, the first chapter sketches in the classical heritage of ancient Greece and Rome, which was the ultimate source of the intellectual currents that so powerfully linked the ancients and the moderns and provided the seedbed for the Scientific Revolution.

Early modern science connected with this classical tradition in another important way. Science, or what, at the time, was called natural philosophy, continued to be understood by many great thinkers, including Paracelsus, Bruno, Bacon, Boyle, Gassendi, Descartes, Hobbes, and Wilkins, as the ancients had done. That is, the moderns, like the ancients, thought of science as laying the foundations for a larger philosophical enterprise, the attempt to arrive at an understanding of humanity and nature, of physics *and* ethics. This, it was believed, would help human beings to achieve more control over both the world and themselves and to realize a higher level of material *and* moral culture. This project, which Descartes called "the search for wisdom," is neglected, if not ignored, in other general accounts of the Scientific Revolution but is nonetheless essential for understanding it, and so it will play a big part here (Descartes 1968, 171–88).

Finally, modern science emerged out of a world caught in the social and religious crisis of the Reformation and that extension of the Reformation known as the English Revolution, and so I have explored how science developed *in part* in response to those profoundly changing times. In particular, I show that, throughout the Scientific Revolution, science and religion were intimately connected. Scientists (or natural philosophers) interpreted their work in religious terms and saw it as contributing to the achievement of their religious goals. Such behavior represents yet another aspect of their "search for wisdom." Religious commitments informed this search and gave it meaning, purpose, and direction. While I do not claim that this contextual approach to the Scientific Revolution lays bare the elusive causes of scientific change, at the very least, it throws considerable light on the motives for doing science, and these, in turn, have considerable bearing on causes and processes, the whys and hows in our story. The contextual approach adopted here also helps to integrate the Scientific Revolution into the general history of early modern Europe and makes this book suitable for use in survey courses devoted to that period or later periods, as well as in history of science courses.

During the last century, the historical study of the origins of modern science has been subject to a variety of approaches. I wish to discuss briefly the approach adopted here by locating it within the historiography devoted to the Scientific Revolution. One hundred years ago, the topic was sometimes treated as an episode in what one historian called "the warfare of science with theology in Christendom" (White 1896). But, in this century, a reversal has set in, and religion is now generally seen as having had a positive, rather than negative, effect on the development of science. The birth of modern science has been shaped by religion in two fundamental ways. First, historians have argued that the great thinkers who made the Scientific Revolution were themselves deeply devout believers whose theological views and religious commitments influenced their scientific thinking (Burtt 1954; Koyré 1968; Lenoble 1943). Second, under the impact of Max Weber's classic *The Protestant Ethic and the Spirit of Capitalism* (1888), other historians have brought the sociology of religion to bear on the history of science. In particular, Robert Merton, in 1938, extended the Weber thesis that the psychology of Calvinism was conducive to the growth of capitalist behavior patterns and argued that English Calvinism or Puritanism led to the development of both capitalism and science in seventeenth-century England. To put it reductively, Merton claimed that Puritans qua Puritans were both inclined to think capitalistically and drawn to do science; a common religious psychology, in other words, had these dual effects (Merton 1970, 55–136).

Since the Second World War, historians of science have disagreed along a spectrum of opinion as to how to properly study science in historical terms. At one extreme are the so-called internalists who argue that science should be treated in terms of the "internal" development of scientific ideas and practices isolated from "external" historical factors, such as religion, politics, or social processes, which are seen by extreme internalists to have had little or no effect on science itself. The history of science, in this view, is the story of scientific problem solving or, in more exalted terms, the search for knowledge about the natural world, unaffected by the wider human world of nonscientific ideas and events.

At the other extreme are the so-called externalists for whom science is regularly, and even crucially, shaped by the context in which it is carried on. This context can be understood to include everything from the laboratories in which scientific research is conducted to the larger society in which scientists live. Most historians of science fall somewhere in the middle, right or left of center and between the two extremes, in this ongoing debate between internalists and externalists.

This book does not adopt an extreme position in this debate but takes the view that science is the product both of internal problem solving and the logic of the search for knowledge, on the one hand, and of the contexts in which research is conducted and in which the scientists actually live and think, on the other. In the case of the major thinkers discussed in what follows, their contributions to science are noted, but the larger intellectual and social framework in which they worked and lived, and its effect on their scientific thinking, are also remarked upon. How could it be otherwise, when, as so many historians have taught us to see, the thinkers who are treated here were motivated by ideas and commitments that were not only scientific but religious, philosophical, and ideological, and they interpreted their scientific work in the light of these nonscientific, so-called external factors? The contextual meaning that scientists gave to the practice, results, and goals of their science will bulk large in our story.

This contextual approach is particularly important for understanding the development of science in seventeenth-century France and England, the two main centers of the Scientific Revolution, to which most of this book is devoted. In both countries, we shall see science evolving in response to new scientific discoveries and theories and to the technical problems they raised that called for scientific solutions. But the problems the scientists faced were not all technical, mathematical, and experimental. They also addressed religious, moral, social, and political issues and thought of their science as laying the groundwork for helping to resolve them as well.

In both France and England, all the leading natural philosophers, as they called themselves, with the possible exception of Thomas Hobbes, were religious believers, and, in most cases, devout ones at that. As such, they were dedicated to defeating heresy and preserving and advancing church and faith. They were also dedicated to developing a science that would provide the intellectual foundations for the achievement of their religious goals. But their sense of mission was not restricted to religious matters. They were often social and political idealists who believed science could (and should) be enlisted to build a new kind of state and to produce more knowledgeable *and* virtuous subjects and citizens, fit for realizing the goals of their commonwealth. "Truth prints goodness," Bacon said, which aptly sums up my point (1955, 215–6). Seventeenth-century natural philosophers believed that their new science should lead not only to "the advancement of learning" but to moral reformation and social regeneration as well. What follows will explore these religious and ideological themes in some detail. It is enough, for the moment, to have made the case for a history of science that excludes neither the internalist nor the externalist side of the story. In fact, on this nonexclusionary, integrative view, it is sometimes difficult, if not impossible, to tell the two sides apart.

Science in early modern western Europe was not just about the conquest of nature but about mastering both fallen humanity and stubborn nature and then using this *dual* conquest to raise the level of civilized life, to build a new material *and* moral order. It is a story of deep significance in terms of both its surpassing intellectual and technical achievements and its broader religious and ideological aspirations.

Quotations from primary sources have been modernized for quick comprehension. The birth and death dates of thinkers have been given in parentheses at the first, or early, mention of their names. For rulers, the parenthetical dates indicate the years of their reign.

My thanks to the series editors for asking me to write this book, and especially to Spencer Weart for his careful reading of successive drafts, which has made it more readable and accurate.

The Classical Legacy

THE ANCIENT GREEKS bequeathed to the West the idea of science, namely, that the universe is a cosmos, an ordered and even harmonious system of nature, and that it can be known through human reason and the senses. This contribution was so fundamental to the whole history of science, and especially to the emergence of modern science in the sixteenth and seventeenth centuries, that this chapter will be devoted to describing many of the salient features of this ancient heritage.

From the sixth and fifth centuries B.C.E. came the speculations of the earliest Greek philosophers as to what the world is made of. For Thales (ca. 625–545 B.C.E.), it is made of water; for Anaximander (610?–546? B.C.E.), it is the Boundless, an undifferentiated stuff, infinite and eternal. For Thales, moreover, the world is made by a transcendent god; but, for Anaximander, god and the world are one. Leucippus (fl. 440 B.C.E.) and Democritus (fl. 410 B.C.E.), the first atomists, postulated a world consisting of an infinity of atoms moving in an infinite void. Particularly important were the ideas of Pythagoras (ca. 572–ca. 497 B.C.E.) and his school. For Thales, Anaximander, and the atomists, matter is nature's basic building block. But Pythagoras and his followers looked for order not in the stuff itself but in numbers. According to Aristotle, they held that all things are numbers or that things imitate numbers. It was a revolutionary step because it meant that, for the Pythagoreans, the key to explaining the order of nature lay in mathematics, especially geometry. Speculation as to what the primitive matter is like was rendered unimportant; the essential nature of things lies in their geometrical structure or form (Lloyd 1979, 145–6; Collingwood 1965, 51–2).

The most influential Greek philosophers were Plato (427–347 B.C.E.) and his student, Aristotle (384–322 B.C.E.). Their work formed the backbone of medieval philosophy (pp. 14–17), and much of the thinking associated with the Scientific Revolution in early modern times represents a dialogue with these two ancients. Plato may have visited a Pythagorean community in his travels in Sicily and southern Italy in the early fifth century, and Aristotle (who should have known) called him a Pythagorean. Plato argued that his rational god, the Demiurge, makes the world by geometrizing primitive, inchoate matter and shaping it into the four elements—air, earth, fire, and water (first postulated by the philosopher Empedocles, fl. 450 B.C.E.). For Plato, following Pythagoras, material particles can be reduced to the elements constituted as geometrical figures, thus opening the way to explaining nature by mathematizing it (Collingwood 1965, 73).

As for Plato's cosmos, the finished product of the Demiurge's creative act, it is finite, bounded by a celestial sphere with a stationary earth at its center. The sun, moon, and planets move around the celestial sphere at different speeds (the sun revolves around the earth once in a year; the moon, once in a month). The celestial sphere itself rotates on its axis once a day, and attached to it are the fixed stars, which thus rotate with it. The world fashioned by the Demiurge is a living creature and possesses a soul responsible for maintaining the cosmic order and the regularity of its motions. As such, the world-soul is a divinity inferior only to its creator, the Demiurge itself. The planets and fixed stars also partake of divinity and help to account for the universal harmony. The heavenly bodies move in uniform circular motions because such motion is perfect and rational and, hence, the only kind worthy of the heavens' divine status (Lindberg 1992, 43).

But this is only half the story of Plato's cosmology. The Demiurge shapes the world and appoints lesser gods than he to maintain its order. He is not a creator *ex nihilo* (out of nothing), like the Christian God, but an agent, and a necessary one. His job is to fashion an imperfect world according to the pattern established in another entirely separate, transcendent world of perfect forms. For, according to Plato, there are two levels of existence, or kinds of being, perfect and imperfect, changeless and transitory. The higher level is the real world of forms or essences, while this lower level we live in and experience is no more than a pale copy. The perfect forms offer a nonspatial, immaterial, eternal, purely intelligible template from which this perceptible world of material objects, ever subject to change, can be drawn. For every such object, there is a form or principle from which it derives its structure and to which it more or less conforms. But, for all their perfection, the pure forms cannot make copies of themselves. They are standards, not agencies. Hence, we must look elsewhere for the active source of movement and life in the world, namely, to the Demiurge, who thus becomes the effective world-maker by copying the forms and imposing geometrical structure on the primitive chaos (Collingwood 1965, 76).

Aristotle's system has much in common with Plato's. For both, nature is alive and eternal, and material objects are actuated by their immaterial forms. But, at this point, Plato and Aristotle part company. Plato draws a distinction between the perfect immaterial forms and the Demiurge, who models the world, by his creative act of will, to that preexisting pattern. But, for Aristotle, the world-process is exactly what Plato said it could not be, namely, a self-causing and self-existing process. In other words, Aristotle eliminated the need for the Demiurge by giving his role to the form of the material object itself. For example, as the historian R. G. Collingwood explains:

> The seed grows only because it wants to become a plant. . . . We can use these words "want" or "desire" because although the plant has no intellect or mind and cannot conceive the form in question, it has a soul . . . and therefore has wants or desires, although it does not know what it wants. The form is the object of these desires: in Aristotle's own words, it is not itself in motion (for it is not a material thing and therefore of course cannot be in motion) but it causes motion in other things by being an object of desire. (1965, 84–5)

Nature, for Aristotle, represents a hierarchy of ends, and each order of beings must have an end of its own. At the top of this hierarchy sits the Unmoved Mover, identical with the forms and completely absorbed in contemplating them. This contemplative activity is the highest and best possible and inspires the whole of nature with desire for the Unmoved Mover and a drive toward reproducing it, everything in its degree, and to the best of its power (Collingwood 1965, 82–5, 87, 89).

Aristotle's universe is made up of substances that are, in turn, composed of matter and form. It is the form of a substance that establishes its nature and therefore its behavior. The universe, bounded by a perfect sphere, is finite and eternal. Unlike Plato's universe, which is homogeneous, Aristotle's is binary, that is, divided into two regions, the superlunary (above the moon) and the sublunary (below the moon).

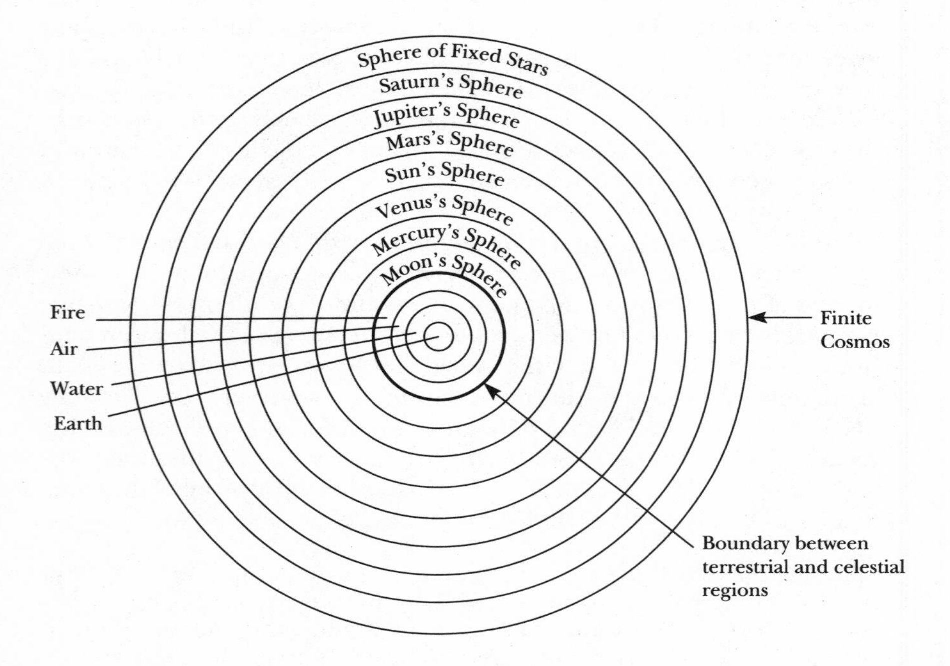

FIGURE 1 The Aristotelian Cosmos

The substances found in the two regions, and their behaviors, are correspondingly different. For Aristotle, the earth is at rest at the center of the world, and the region below the moon is made up of the four elements (Fig. 1). Air and fire are light (though fire is lighter) and will naturally rise; earth and water are heavy (though earth is heavier) and will descend to the center of the world because it is their nature to do so. Terrestrial objects seek their natural place in the sublunary region and have a natural motion, which is to rise (if light) or to fall (if heavy).

Heavy objects may also have unnatural or "violent" motion—for instance, an arrow shot from a bow. In such a case, the arrow is said to defy its natural motion, and its forward thrust requires an explanation: The air displaced by the moving arrow is thought to swirl around behind and push the arrow forward. The speed of falling terrestrial objects is determined by both their weight and the resistance of the mediums through which they pass. Aristotle held that heavier objects fall faster than lighter ones and that the speed of any object varies indirectly with the density of the surrounding medium. Terrestrial motion is conceived of, by Aristotle, as one of four kinds of change that sublunary substances may undergo. (Celestial things are changeless.) The other three types of change are

1) generation and corruption (for example, birth and death),
2) alteration of quality (for example, of color or texture), and
3) alteration of quantity (for example, of size or number).

For Aristotle, the terrestrial region is a setting for incessant change of these four types. But the heavens are altogether different. According to Aristotle, it is a matter of observation that the heavens never change. They must therefore be composed of an incorruptible fifth element, called ether, which fills the entire celestial region (Fig. 1). Following the complex scheme worked out by Eudoxus of Cnidus (ca. 390–ca. 337 B.C.E.), Aristotle held that the planets and stars are attached to perfect concentric spheres that revolve with the most perfect of movements, that is, eternally unvarying circular motion, around a stationary earth. The cause of this motion is the desire of the spheres to imitate the perfection of the Unmoved Mover. Here is a particularly clear case of Aristotle's doctrine that the universe is made up of substances activated by their forms with the desire for self-actualization. The heavens—concentric spheres, planets, and stars—are divine. "We are inclined," Aristotle said, "to think of the stars as mere bodies . . . ; whereas we ought to think of them as partaking of life and initiative," (quoted in Brooke 1991, 119) just like plants and animals, only more so. The celestial bodies are moving divinities. "Beyond them, incorporeal and outside the universe, is the primary divinity, the changeless originator of all change," the Unmoved Mover (Barnes 1982, 64).

Aristotle's authority in cosmology was never absolute in the ancient world. But it was enough to overcome rival theories and to provide a foundation for further development in astronomical theory. Aristarchus of Samos (ca. 310–230 B.C.E.) put a stationary sun at the center of the universe with the earth revolving around it. The sphere of the fixed stars remains unmoved in his scheme. To account for the fact that we see them in the same place every night (and for sunrise and sunset), Aristarchus made the earth rotate on its axis once a day (diurnal rotation). However, this heliocentric model was adopted by only one other ancient astronomer, Seleucus, who worked in the second century B.C.E.

Aristarchus's heliocentrism was defeated for a combination of reasons. For most Greeks, the earth was considered sacred, and geocentrism was a mark of that sanctity. At least one philosopher charged Aristarchus with impiety for putting the earth in motion. A stationary earth could also be defended on scientific grounds. Aristotelian physics necessitated such a universe: Heavy objects naturally fall toward the center of the world and come to rest there. The earth, the heaviest object, must therefore be at rest at the center toward which all heavy things fall. If the earth rotated on its axis or revolved on its sphere, heavy objects thrown into the air or things suspended above the earth (clouds, for instance) would be left behind—far behind, because of the tremendous speed the earth would have to attain to cover such great distances in the time allotted (twenty-four hours in the case of diurnal rotation). And heavy things resting on the surface of the earth, but not securely tied down, would also be left behind because the earth is very heavy, and speed, according to Aristotle, varies directly with weight. Another argument against Aristarchus was the fact that the stars were always observed to be in the same position relative to the earth at all times of the day and the year, which suggested that the earth does not move at all. Aristarchus countered by claiming that this apparent effect is due to the fact that the stars are infinitely distant from the earth, but his explanation was not convincing to those persuaded by Plato and Aristotle that the universe is finite (Lindberg 1992, 97–8).

Following in the footsteps of earlier observers of the skies, Claudius Ptolemy (fl. 150 C.E.) summed up the development of astronomy in the ancient world by constructing the most elegant mathematical model of the motions of the planets. Ptolemy himself was a convinced Aristotelian and devoted himself to devising a model that, complex though it was, would uphold the requirement of uniform circular motion in the heavens. He committed himself, in other words, to "saving the phenomena," that is, to preserving Aristotle's binary universe of perfect heavens swinging forever around a stationary earth.

He adopted three devices. From Appolonius of Perga (fl. 210 B.C.E.), he borrowed the twin models of the eccentric circle and the epicycle on a deferent. The eccentric model (Fig. 2) represents the circular path of a planet (P) whose center (C) is not the earth (E), which thus becomes eccentric. The epicycle on a deferent model (Fig. 3) is the circular path of a planet (P) whose center (C) is represented as a point on a larger circle, the deferent (D), centered on the earth (E).

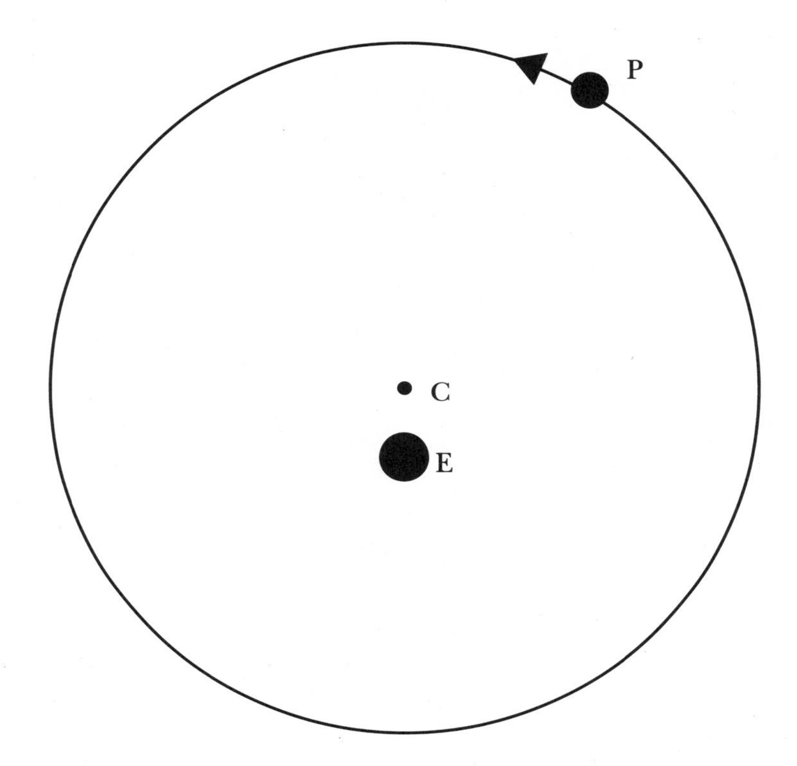

FIGURE 2 Ptolemy's Eccentric Circle

The third model of planetary motion, the equant (Fig. 4), was Ptolemy's own invention. To understand the equant, we start with the eccentric model, and then we add the equant point (EP), a noncentral point as off center in one direction as the earth (E) is in the other. Following Ptolemy, we can then say that the planet (P), shown moving in a circle, sweeps out equal angles in equal times as measured not from the center (C) but from the imaginary equant point. Ptolemy's purpose in deploying these three devices was to square the *observed* irregularity of planetary motion with the *theoretical* requirement of uniform circular motion. The Scientific Revolution would begin when Copernicus broke from Ptolemy's picture of the universe, not because the fit he had established between theory and observations was not convincing at all and so his theory had to be discarded, but, paradoxically enough, because

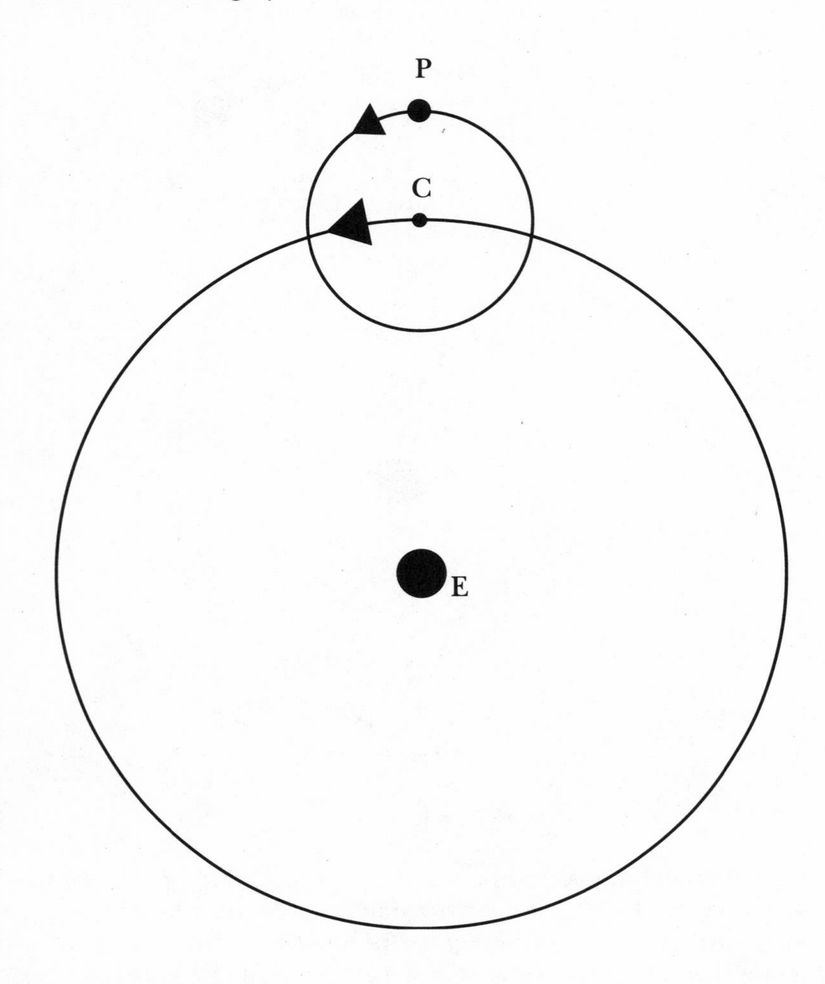

FIGURE 3 Ptolemy's Epicycle on a Deferent Circle

the fit was not perfect enough, and so his theory had to be adjusted. In particular, the offending equant, which Copernicus thought a blemish on so noble a picture, had to be removed. The deep and lasting commitment to uniform, circular motion was to survive and even promote the first stages in the Scientific Revolution.

Ptolemy's achievement was remarkable and had multiple effects. It preserved the binary system of Aristotle's universe. It enabled astrono-

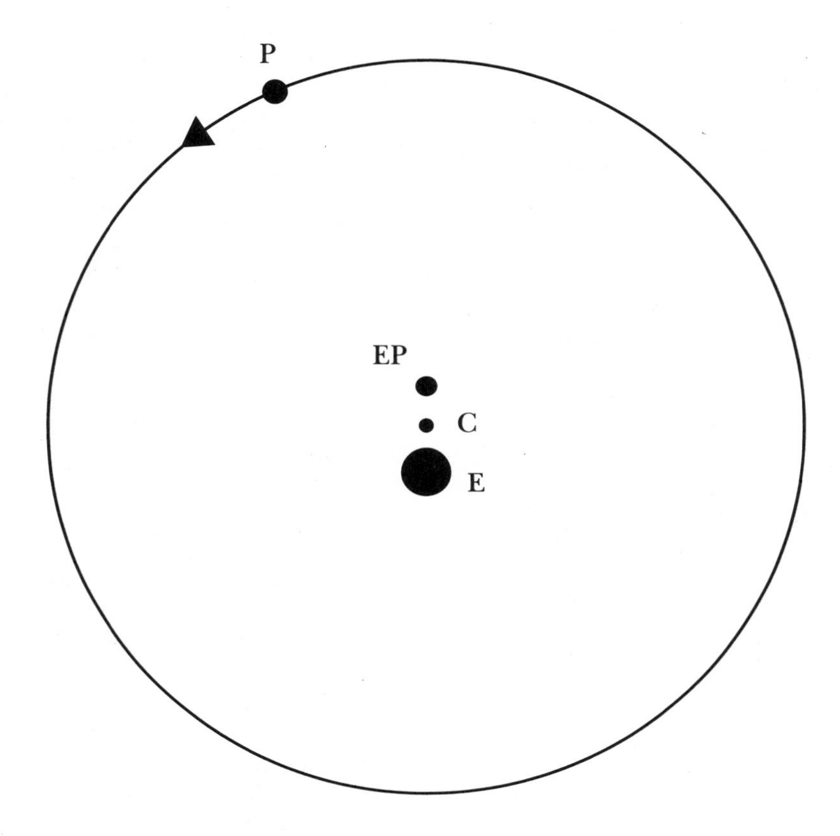

FIGURE 4 Ptolemy's Equant Circle

mers to make reliable predictions of future planetary positions, and, on that basis (and no less important to Ptolemy), it supposedly enabled astrologers to predict future events on earth. For Ptolemy, as earlier for the Pythagoreans, the mathematical study of the heavens also had an ethical dimension: "Astronomy teaches man to regulate his life in imitation of the order, constancy, and tranquillity of the celestial motions" (quoted in Taub 1993, 5). Finally, Ptolemy's was the model that was handed down and accepted through the centuries as offering the true picture of the heavens, until it was eventually challenged and gradually undermined in the sixteenth and early seventeenth centuries, almost fifteen hundred years after it was first put forward in Ptolemy's great work, the *Almagest* (150 C.E.) (Lindberg 1992, 99; Lloyd 1979, 200).

Two other schools of philosophy rose up in ancient Greece, the Epicureans and the Stoics, both of which would constitute important ingredients in the Scientific Revolution of the sixteenth and seventeenth centuries. Epicurus (341–270 B.C.E.) revived, with major differences, the atomism of Democritus and Leucippus. The universe, he said, is uncreated and boundless in both space and time. The infinite void contains an infinite number of worlds, some like ours and some not, all of which are composed of moving atoms. Each such world comes into existence and eventually is dissolved back into its atomic constituents, from which, in an endless cycle, new worlds are made (Fig. 5). Everything is formed out of the impact of matter in motion. There are gods, but they are also material and have nothing to do with creating or maintaining the world process. The Epicureans held that "the flaws in the observed character of the world confirm the atomic theory, and undermine belief in gods who care about the world" (Irwin 1989, 155).

In such a system, there is neither the purpose that Aristotle attributed to the forms nor the design that Plato attributed to the Demiurge. But the atoms occasionally undergo a random swerve or uncaused motion, which prevents a completely mechanically determined universe and provides a basis for an argument of human free will. Every choice a person makes constitutes an instance of random atomic swerve. One's duty, then, is to make informed decisions that will allow the individual to lead a happy life. Epicurus was a hedonist, and a happy life was a matter for hedonic calculation. This did not mean, for him, that we should search for greater and greater satisfaction or ever more intense delight. On the contrary, in an imperfect mechanistic world devoid of purpose or design, where pain and misery are so common, the object is to achieve a balance of pleasure over pain. Basic needs for food, shelter, and health must be met, but much more than that might prove excessive and bring more pain than pleasure. As Cicero, interpreting Epicurus, said: "For who does not see that it is need that gives savor to everything." The goal to be aimed for is freedom from hardship, untroubledness, self-sufficiency, tranquillity, all of which require self-mastery. By regulating his desires, the Epicurean strives to minimize his dependence on external fortune, therefore valuing the results of temperance, the very virtue also closely associated with Plato and Aristotle (Lovejoy and Boas 1980, 154; Irwin 1989, 159; Long 1974).

Epicureanism had some considerable success in the ancient world,

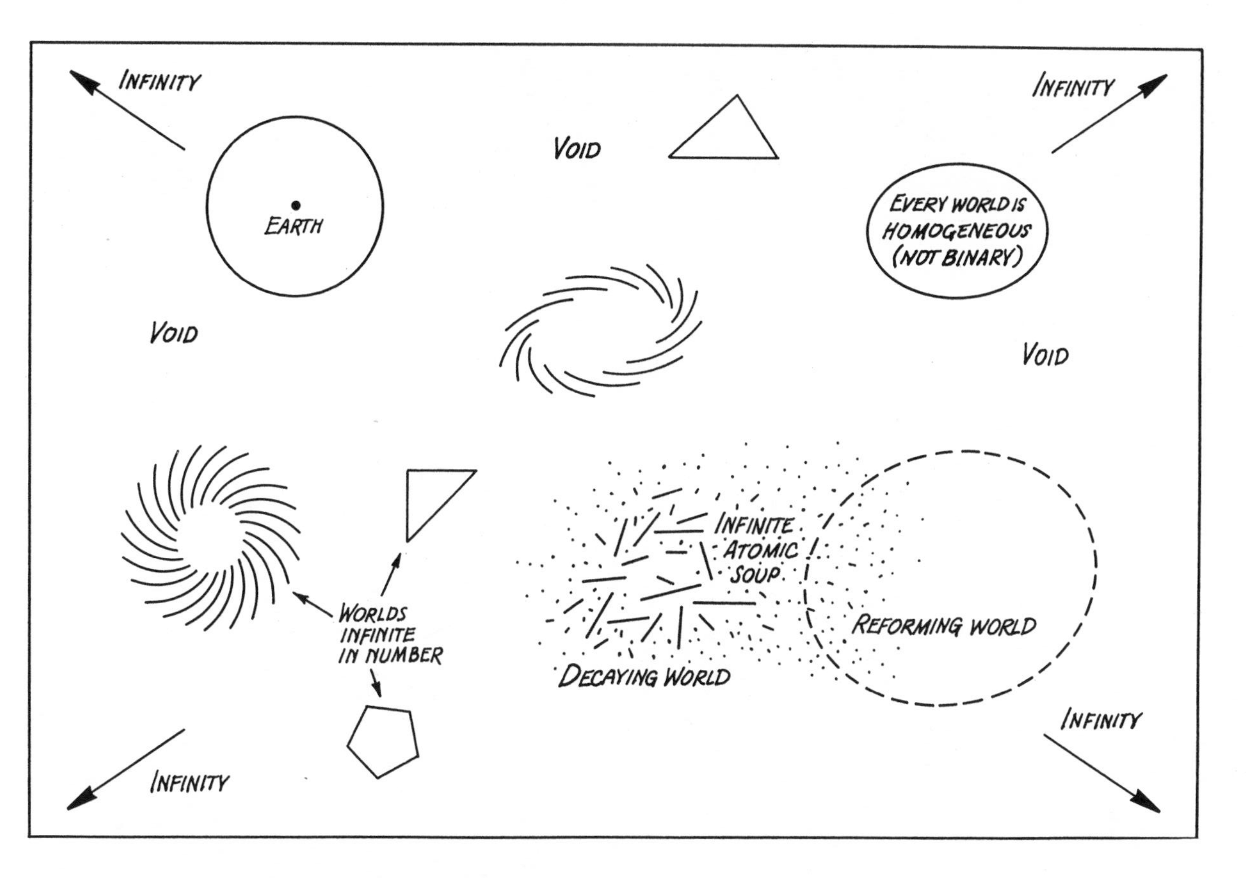

FIGURE 5 The Epicurean Universe. (Illustration by Robert Trondsen.)

but, with the rise of Christianity, it was condemned for obvious reasons; both its cosmology and its ethics could only be deemed heretical. It was not until the fifteenth and sixteenth centuries that Epicureanism began to be discussed and the key texts disseminated, of which the most important was the great poem *De Rerum Natura* (*On the Nature of Things*) by Lucretius (ca. 96–ca. 55 B.C.E.). As we shall see, Epicureanism was to play a major role in both ethical thought and the revival of atomism in the seventeenth century (Allen 1944; Sailor 1964).

According to the Stoics, the universe consists of a cosmos, our geocentric world, surrounded by an infinite extracosmic void. The cosmos is made of continuous matter, a plenum, not binary like Aristotle's, but homogeneous. The Stoic void and cosmos are uncreated and eternal, but the cosmos, our world of stars and planets, is subject to an eternal cosmic cycle of birth, expansion, conflagration, contraction, and regeneration (Fig. 6). Stoicism, like Epicureanism, represented a kind of materialism, but where Epicurus denied any plan or purpose to his universe, the Stoic cosmos was pervaded by a subtle matter, or divine breath, which functioned as a world-soul, directing and organizing nature. Epicurus was a mechanistic materialist; in contrast, Stoic materialism was organic and vitalistic, if not sometimes even pantheistic. And there was another fundamental difference. Where Epicurus introduced randomness into nature through the doctrine of the swerve, the Stoics were rigidly deterministic.

If the cosmos is the product of an immanent and rational divine agency, as the Stoics taught, it is incumbent on all human beings to use their reason to discover and obey the natural law that governs the world, that is, to live according to nature, as Aristotle had also said. The result of such effort would be individual moral progress, the increasing strength to master internal drives and to overcome external circumstances and thus to achieve detachment, imperturbability, and happiness.

The founder of the Stoic school, Zeno (ca. 333–262 B.C.E.) of Cyprus, found followers among Greeks of the third and second centuries B.C.E. But, because of their unremitting naturalism and materialism, the early Stoic writings of the Greeks ceased to be copied, especially by Christians, and were effectively lost. The knowledge of Stoicism that spread in the Middle Ages and in the sixteenth and seventeenth centuries was, therefore, secondhand. The two most important ancient sources of this knowledge were the Latin writers Cicero (106–43 B.C.E.) and Seneca (4 B.C.E?–65 C.E.). Among medieval Christian thinkers, Stoic

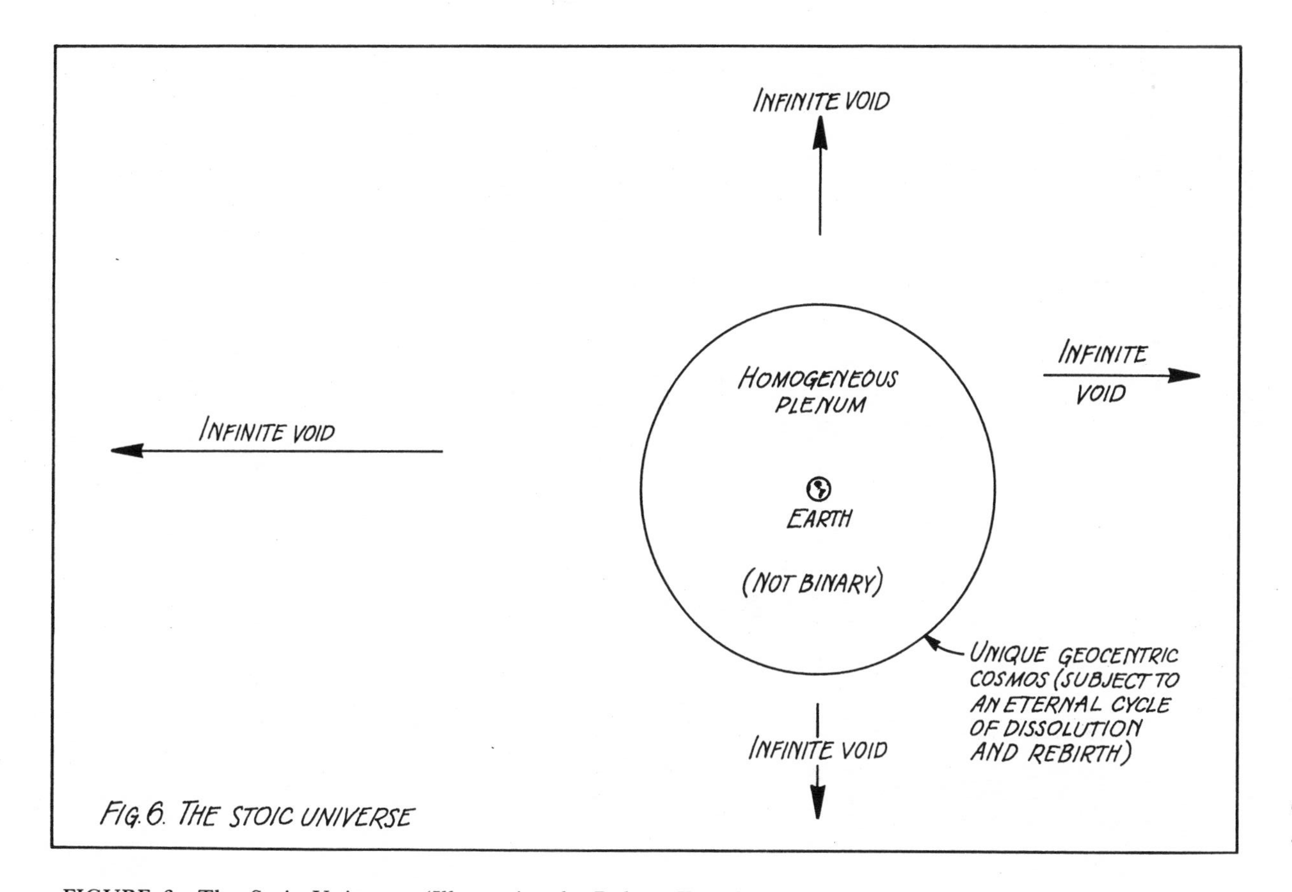

FIGURE 6 The Stoic Universe. (Illustration by Robert Trondsen.)

vitalistic materialism was still anathema, but Stoic ethics remained attractive, especially the duty of the wise individual to live in accord with nature and to obey its laws—which were thought by Christian philosophers to be identical with divine reason and will. In general, Stoic ethics could easily be assimilated to Christian teaching (Lapidge 1988, 97). In the sixteenth and seventeenth centuries, the ancient Stoic legacy of cosmology and cosmogony was revived, as we shall see, with important consequences for the Scientific Revolution (Lapidge 1988; Long 1974).

But the ancient tradition that turned out to be the most successful in the Middle Ages, and that remained vital up to the middle of the seventeenth century, was that of Aristotle and Ptolemy. The transmission of this tradition across a thousand years proved circuitous. The prophet Mohammed (ca. 570–632) founded the new religion of Islam in the Arab cities of Mecca and Medina. His Muslim successors, inspired by a faith that preached holy war, went on to conquer Persia, the Near East, North Africa, and Spain by the middle of the eighth century. Over the next three centuries, Muslim scholars, working in Baghdad, Cairo, and Spain, preserved Greek scientific manuscripts and translated them into Arabic. The manuscripts in Spain became available to Christian scholars as Muslim libraries, especially the great library at Toledo, fell into Christian hands during the Reconquest in the late eleventh and early twelfth centuries. Ancient science also trickled into Italy through contact with Muslim trading posts in Sicily. From the twelfth century on, Greek science and its Arabic commentaries were translated into Latin by Christians. Ptolemy's *Almagest* and most of the works of Aristotle were available in Latin translation by 1200 and grew more influential in the next several decades. During the second half of the thirteenth century, lectures on Aristotle became an important part of the curriculum of the medieval universities, the first of which, Bologna, was founded about 1150 (Lindberg 1992, 166–82, 203–13).

The Muslim Arabs also transmitted and bequeathed to the Christian West the Hindu-Arabic number system. A treatise on the subject was one of the first Arabic texts to be translated in the twelfth century. The Muslims also developed algebra, itself a word derived from Arabic. The use of algebra, derived from Arab sources, spread throughout western Europe during the Renaissance, and several Italian mathematicians, notably Niccolò Tartaglia (ca. 1501–57), contributed to its further development. Girolamo Cardano (1501–76) summed up these advances in his *Ars Magna* (*The Great Art*), published in 1545. A century later, Pierre de Fermat (1601–65) and René Descartes (in *La Géométrie,* 1637) went on to show how algebra could be used to solve geometrical problems (Hall 1962, 227–32).

During the thirteenth century, as Christian scholars in the West assimilated the pagan philosophy of Aristotle, his work was cleansed of its paganism and made to conform to orthodox Christian doctrine. In particular, Aristotle's eternal universe gave way to a created one, his Unmoved Mover to a providential God, his souls inseparable from bodies to the immortality of the soul (that is, the doctrine that when the body dies, the soul survives to receive its just deserts in the afterlife). Chiefly responsible for thus Christianizing Aristotle were two thinkers, Albert the Great (1200–80) and Thomas Aquinas (ca. 1224–74). For them, there could be no conflict between Aristotelian philosophy and Christian theology; rather, the former served as handmaiden to the latter. God and nature, they said, can be understood both by reason (philosophy) and by revelation (theology). Any case of apparent conflict between the two is the result not of true philosophy but of erroneous thinking. There are, however, truths of revelation that cannot be explained by reason—the biblical miracles, for instance. These, Albert and Aquinas claimed, must be accepted on faith, without rational assistance (Lindberg 1992, 215–23, 228–34).

There were also radical Aristotelians who pressed the case for the eternality of the world or the immortality of the soul, and they were roundly condemned by church authorities. Thus, at the same time that Aristotle was being accommodated to Christian theology, the church had to fight a rearguard action against Aristotelian heresy. In the end, it was the reconcilers and Christianizers who won: By the fourteenth century, Aristotelianism, made safe for the church, was officially taught in western European universities, and any dissent was made difficult and dangerous (Lindberg 1992, 234–44).

In cosmology, the homogeneous cosmos of Platonists and Stoics, which had been taught up through the twelfth century, gave way in the thirteenth century to the binary cosmos of Aristotle and Ptolemy. But the Aristotelian idea of a finite universe came into conflict with the Christian doctrine of divine omnipotence. The issue was resolved by maintaining that God has the power to create other worlds besides ours and that therefore there may, or must, exist an infinite void space to hold these potential worlds. Thomas Bradwardine (1290–1349) identified this void space with God's omnipresence and argued that, since God is infinite, extracosmic void space must likewise be infinite. Christian doctrine shaped the issue, but there was also Stoic influence. "The notion of [an] extracosmic void came to the West with a Stoic pedigree" (Lindberg 1992, 248). Theological pressures thus forced a Stoic modification of Aristotelian cosmology, a modification that would play a major role in early modern cosmological speculations. With this important exception, the medieval cosmos was that bequeathed by Aristotle and Ptolemy—a set of tightly nested concentric spheres without any space between, but nevertheless moving with frictionless, uniform motion. The spheres were now visualized as being thick enough to accommodate the eccentric deferents and epicycles of Ptolemaic astronomy (Lapidge 1988; Lindberg 1992, 245–52).

During the thirteenth century, astrology (that is, the study of the heavens for the purpose of predicting the future) became well established and would flourish until the late seventeenth century. Astrology derived its theory from a mixture of Platonic and Aristotelian sources. From Plato came two supports: (1) the notion of planetary deities affecting the sublunary realm and (2) the unity of the cosmos and, in particular, the close link between the macrocosm and the microcosm (the world and the individual human being). Aristotle, for his part, had argued that terrestrial change derives from celestial motions. Until the seventeenth century, most philosophers, though not all, took for granted the causal power of the heavens over the earth; astrology was the science that made sense of this and allowed for prognostication (Lindberg 1992, 274–80; Schmitt and Skinner 1988, 286–7). Astrology was also thought to be fundamental to the theory and practice of medicine. When deciding on a treatment, it was important for the university-trained physician to take celestial influences into account.

Late medieval and early modern Aristotelianism was known as Scholasticism, the doctrine of "the schools" or universities, and indeed made up the bulk of the university curriculum. Universities were the main centers of science instruction. They were also centers of scientific research, though royal courts and aristocratic households served that function as well. The relation of the universities to the new thinking associated with the Scientific Revolution was complex. Much of the new science of the sixteenth and seventeenth centuries attacked Scholasticism as false and useless and, over the long run, destroyed and replaced it. But the universities, though dominated by Scholastic teaching, were not rigid and unchanging. Instead, they sometimes fostered changes that encouraged new thinking and that, no doubt, led in time to the downfall of Scholasticism itself. Many of the most important thinkers of the Scientific Revolution were university educated, and the instruction they received often prepared them to go on to do independent thinking and to achieve breakthroughs. Galileo and Newton's student notebooks make this abundantly clear. Yet Scholasticism was not always the hurdle that had to be overcome. In the field of anatomy, for example, the new revolutionary thinking, as we shall see, occurred within what remained, in many respects, a basically Aristotelian framework (Gascoigne 1990; Hall 1962, 238–41).

CONTROL OF NATURE 2

Cosmological Renewal and Corrosive Doubt

ARISTOTLE AND PTOLEMY held sway among educated people until their authority in physics and astronomy was overthrown in the course of the seventeenth century. But a number of developments in the fifteenth and sixteenth centuries are important for understanding that later process.

Renaissance Humanism and the Print Revolution

Works of ancient science and philosophy continued to be discovered and translated into Latin and the vernacular languages. Much of this activity was undertaken in the fifteenth- and sixteenth-century Italian city-states by the so-called Renaissance humanists. Humanism was a movement inspired by the example of ancient Greece and Rome and sought to recover as much of this lost classical culture as possible, including its rich legacy of natural philosophy and mathematics. The humanists believed that, once this ancient heritage had been retrieved, it might be put to work to provide a new kind of education for the elite, to set higher intellectual and artistic standards, and to raise the level of civilized life. Just as the translation of Aristotle's works had a dramatic impact on medieval intellectual life, so the humanists' recovery and interpretation of philosophical works by other ancient thinkers would have a profound effect in subsequent centuries.

The printing press was invented in the mid–fifteenth century (1445–50), and the newly translated texts could now be published and, once in print, spread much faster and cheaper than before. There was also a commercial incentive for such dissemination: Printed books made money. The book trade stimulated intellectual life, including scientific thinking. Not only were texts now more accessible to those who could afford and read them, they also became more accurate because errors could be permanently eliminated as they were detected. But in early modern Europe, printing was not an unmixed blessing (Febvre and Martin 1984).

Elite opinion was often divided, or at least ambivalent, on the issue. Printing was an obvious boon; but the authorities also saw the need for censorship, for imposing controls on what was published. Today, ours is a pluralistic society that values new ideas and a variety of opinion, even differences of opinion, at least up to a point. Early modern Europeans were not so tolerant; in fact, they were suspicious of intellectual novelty and variety and sometimes downright hostile to heterodox thinking. For centuries, leaders in church and state had tried to build unity by enforcing Christian orthodoxy and punishing heresy. The motto of the French monarchy aptly sums up the point: "one king, one law, one faith."

The Protestant Reformation and the ensuing religious wars only made things worse, as religious unity broke down and became impossible to recover. The preaching of what was deemed to be false doctrine was a live issue in early modern Europe, made still more worrisome by the possibilities for spreading it through the press. The fear, moreover, was not only of heresy but of the subversion and anarchy to which such thinking might lead. Lutheran doctrines like "Christian liberty" and "the equality of all believers" were at once fundamental to the Protestant Reformation and rife with radical social implications. Indeed, they became rallying cries for popular rebellion. The Catholic Church established an *Index of Prohibited Books* in the mid–sixteenth century, and Protestant authorities also set up machinery of one kind or another for regulating the book trade. All intellectual life, including the development of science, would be affected by this atmosphere, the tension between the power of print and the concern to control both opinion and behavior (Burke 1978; Febvre and Martin 1984, 216–319; Hill 1972; Williams 1992).

The study of Greek atomism was revived when the first printed edition of Lucretius's *On the Nature of Things* appeared in Brescia in 1473; about thirty editions followed by 1600. The other principal source for Epicurean atomism was *Lives of the Philosophers* by Diogenes Laertius (early third century), translated into Latin about 1430 and first published in Basel in 1533 (in Greek). Plato was another Greek philosopher well served during the Italian Renaissance. He had many fifteenth-century translators, whose work culminated in the Latin translations made by the Florentine Marsilio Ficino (1433–99) and published in 1484. There were more than thirty printings of Ficino's translations in the sixteenth century (Schmitt and Skinner 1988, 77–86, 776–91; Copenhaver and Schmitt 1992, 1–18, 35, 145; Sailor 1964).

Humanism had another positive effect on scientific development. University education, as we have seen, was Scholastic, that is, based on Aristotle, and accorded a rather minor role to mathematics. But the Scholastics were not resistant to all change, and the curriculum evolved over time. Between the late fifteenth and early seventeenth centuries, western European universities provided more mathematics professorships and introduced more mathematics instruction into the curriculum than ever before. Several humanistic influences worked in this direction. The new interest in mathematics was sometimes associated with the humanist desire to revive classical learning in all its fullness, including the translation of ancient mathematical works by Euclid (fl. 300 B.C.E.) and Archimedes (287–212 B.C.E.). Sometimes this new interest was associated with the Christian humanism of the Protestant Reformation or the Catholic Counter-Reformation. For instance, Philipp Melanchthon (1497–1560), the Lutheran leader, increased the teaching of both mathematics and astronomy at the University of Wittenberg, Luther's own university. For Melanchthon, the Protestant Reformation and the reform of learning went hand in hand. The Catholic Jesuits, who also valued mathematics in their schools, made the same connection between educational and religious reform. And sometimes humanists encouraged mathematics instruction because of their belief that learning should be useful and practical; skill in mathematics was seen to be applicable to many fields, including astrological prognostication and healing, architecture, navigation, cartography, and gunnery (Gascoigne 1990; Westman 1980).

A combination of such motives led Tartaglia, in 1543, to publish the first Latin translation of the works of Archimedes, one of the finest mathematicians and physicists in the ancient world. This was followed in 1544 and 1558 by more accurate and complete translations, both the products of humanist scholarship. At the end of the sixteenth century, the work of Archimedes was a major influence on Galileo, who referred to him as "the superhuman Archimedes, whose name I never mention without a feeling of awe" (quoted in Hall 1962, 215). The

Scholastic tradition could be, and was, emended by the humanists in a way that fed new scientific thinking. That tradition was not always the enemy of such thinking, even though it was often cast in that role by the polemicists for the new science (Hall 1962, 197–227).

Renaissance Magic

Ficino was a follower of Neoplatonism, a viewpoint that sought to reconcile Platonic philosophy and Christian theology and thereby to promote moral purification and spiritual enlightenment. He was deeply influenced by the Greek cardinal Basil Bessarion (1403–72), for whom "Plato was a precursor of Christ" and the goal of philosophy was to discover and teach a theology, very ancient and pure, that would pacify and inspire humankind (Copenhaver and Schmitt 1992, 142). This same quest motivated Ficino. He found a key to this project in the so-called *Corpus Hermeticum*, a collection of texts forged between 100 and 300 C.E. but accepted by Ficino and others as the works of Hermes Trismegistus, an Egyptian priest who, it was claimed, had lived at the time of Moses. According to tradition, Hermes was the inspired author of "an ancient theology which paralleled and confirmed the revealed truths of scripture," a theology which "reached absolute perfection with the divine Plato" (Copenhaver and Schmitt 1992, 146–7). Ficino translated some of these Hermetic texts, which remained influential for another two centuries. He also wrote *On Arranging One's Life According to the Heavens*, the most influential Renaissance treatment of the theory of magic.

Ficino's magic is based on the Platonic and Stoic idea that the human soul is a microcosm of the world-soul that animates the macrocosm. The soul of the world links the individual human being to the world and links everything in the terrestrial world to the heavens. These linkages allow the magician, through his knowledge of the correspondences between higher and lower things, to operate for his own (or his client's) welfare. Magic can reward its beneficiaries with health, prosperity, and spiritual purity. Certain earthly things—for instance, plants, gems, music, and scents—help in concentrating celestial influences, and the magician knows how to use these things as so many lures or charms to call down the benefits of the stars and planets. Ficino insisted that his magic was natural and innocent rather than demonic, but there was a fine line separating the two kinds of magic (Copenhaver and Schmitt 1992, 143–63; Couliano 1987, 111–43).

Ficino initiated a tradition of Hermetic and magical thought that was carried on by his successors up through the first half of the seventeenth century. The first of these was his friend Giovanni Pico della Mirandola (1463–94), who sought to find the common thread that linked all philosophies, including Cabala (a system of medieval Jewish mysticism). Out of this syncretic approach, Pico hoped to provide a new unifying and healing foundation for Christianity that would overcome sectarian discord. His most controversial claim, for instance, was that "there is no science that gives us more certainty of Christ's divinity than magic and Cabala" (quoted in Copenhaver and Schmitt 1992, 169), an assertion that seemed to subordinate Christian truth to the suspect knowledge of wizards and Jews. Both Pico and Ficino subscribed to an optimistic view of human beings, according to which, at least some gifted individuals possess the capacity to control the forces of nature. For Ficino, man the microcosm can manipulate macrocosmic forces for human ends. Going further, Pico gave man the freedom to shape his own destiny. In Pico's *Oration on the Dignity of Man* (1496), God says to man: "We have made you a creature" such that "you may, as the free and proud shaper of your own being, fashion yourself in the form you may prefer" (Copenhaver and Schmitt 1992, 163–76; Pico 1956, 7). This exalted conception of the capacity of at least some men would continue to inspire natural philosophers and to serve as a motive for doing science throughout the Scientific Revolution.

Paracelsian Magic

One thinker who used, for his own purposes, the magical tradition initiated by Ficino and Pico was Theophrastus Bombastus von Hohenheim (1493–1541), who called himself Paracelsus ("beyond Celsus," an ancient Roman medical writer). He was a Swiss physician who revolted against medical tradition and whose natural philosophy and theology crucially influenced science for more than a century after his death. Responsible for some well-publicized cures, he was appointed city physician in Basel but was eventually forced into exile because of his radical medical views and abusive tactics in putting them forward. He held that alchemy should be directed not to transmuting base metals into gold but to discovering and preparing effective chemical medicines. Toward that end, he championed the use of morphine, lead, mercury, and sulfur in treating diseases and was particularly concerned to treat the occupational diseases of miners, including silicosis (Pagel 1958).

Academic medicine in the Renaissance derived from ancient Greek sources, namely, the texts attributed to Hippocrates of Cos (ca. 460–ca. 370 B.C.E.), Aristotle, and, especially, Galen (129–ca. 210 C.E.), and taught that the body contains four fluids, or humors—blood, phlegm, black bile, and yellow bile. Health results from the proper balance of the humors; illness is the result of imbalance. The task of medicine is to restore the balance in a sick patient through a variety of therapies—diet, drugs, and purgings, including laxatives, bloodlettings, and emetics—to eliminate excess fluids. Galenists also held that there is an astrological element in the origins and treatment of disease. It was a commonplace of Aristotelian natural philosophy that the heavens influence the sublunary regions, including the environment and individual human bodies. From such influence, it was thought, can come outbreaks of plague, for example. Likewise, to attempt a cure, it is important for the physician to administer drugs or for a surgeon to bleed a patient at the right time for favorable celestial influence (Lindberg 1992, 111–31).

Paracelsus spent his career attacking this kind of medicine and did so on the basis of a revolutionary rethinking of the causes and treatments of disease and of the nature and functioning of the universe itself. To the four sublunary elements, Paracelsus added three alchemical principles—sulfur, mercury, and salt—to explain the behavior of matter. Everything in nature is endowed by God with a divine spark, a defining *archeus* that gives it a unique character. Illness, according to Paracelsus, is not the result of humoral imbalance but, rather, of the *archeus* of a disease that works against the *archeus* of a healthy body. The physician's job is to prepare a specific drug whose *archeus* opposes the *archeus* of the patient's disease (Henry 1991, 212–3). Paracelsians often referred to the medicines they dispensed as "specifics," that is, targeted to fight a particular disease in a particular patient.

To find such a cure, Paracelsus relied not on traditional printed authorities, like Galen, but on experience. The wise doctor learns from nature: "In one herb there is more virtue . . . than in all the folios that are read in the high colleges." The discovery and extraction of these "virtues," in order to effect cures, constitutes a form of natural magic, and such Paracelsian magic was to have a wide influence (Schmitt and Skinner 1988, 289–91). The curious healer should consult not only the experts but "old women, gypsies, magicians, wayfarers, and all manner of peasant folk" for medical knowledge. In the search for the right medicine for a particular patient, the physician has God on his side: "For there is no sickness against which some remedy has not been created and established [by God], to drive it out and cure it" (Henry 1991; Paracelsus 1951, 50, 57, 77).

Paracelsus subscribed to the idea of the Fortunate Fall, an idea that would continue to be important in the development of early modern science. According to the Old Testament, God's punishment for Adam's sin was a curse by which man was ever after condemned to a life of hard labor. But Paracelsus translated this tragedy into a cosmic opportunity for humankind. Before the Fall, human beings may have lived innocently and happily but were crucially incomplete: "The knowledge that man needs was not yet in Adam but was given him only when he was expelled from Paradise." This precious new knowledge is what Paracelsus called "the light of nature," by which, through sustained exertion, we may achieve a complete understanding of the universe, not all at once but one step at a time. Here is our true mission: "It is God's will that nothing remain unknown to man as he walks in the light of nature; for all things belonging to nature exist for the sake of man." But there is still a catch; "the light of nature" does not let us off the moral hook because with that light also comes dark

> knowledge of good and evil. God in his benevolence has set before our eyes things that we desire—good wines, fair women, good food, good money. And this is the test: whether we keep ourselves under strict control, or whether we break and exceed the measure of nature.

To help humankind out of this predicament, Paracelsus preached his philosophy of work: "Let us not be idlers or dreamers, but always at work, both physically and spiritually, so that no part of us remains inactive." In particular, Paracelsus recommended that we should spend our energies in the search for natural knowledge, the progressive discovery of nature guided by "the light of nature" (Paracelsus 1951, 103–11, 218).

In this respect, Paracelsus was an early participant in a wider European movement allied to both science and Renaissance humanism. Both scientists and humanists emphasized practical education and the active life. People of all social strata should be trained for useful employment beneficial to both themselves and the commonwealth. Skilled work was seen as a key to moral rejuvenation at both ends of the social spectrum, that is, good for both the gentry and the common people. The resulting social order was often envisaged in utopian terms as a place where not only would everyone be busy and well fed but greed would be banished, where, in other words, there would be neither too much nor too little but just enough for all. Implicit in this philosophy of full employment was an attack on the traditional hierarchical thinking that despised manual labor and the mechanical arts and exalted a life of leisure, whether spent in gentry pursuits or in scholarly contemplation (Rossi 1970; Davis 1981, 11–104, 299–367; Todd 1987).

As with so much else, Paracelsus turned traditional astrology upside down. The conventional view was based on the ancient Greek notion of the superiority of the heavens and their influence on the inferior sublunary region. For Paracelsus, too, the stars are the source of enormous divine power and wisdom to be tapped by human beings. But there is no need to look to the heavens for these resources because, ever since the biblical Fall, the influence of the heavens has been preempted by the light of nature with which God has endowed humankind. The stars are "active in us" and will, if we let them, teach us "all wisdom." Those who find this power within them become "holy men . . . who serve the forces of nature, and they are called magi." Here is Paracelsus's reversal of astrological tradition: "The stars are subject to" such a magician; "they must follow him, and not he them." Any man, on the other hand, who does not know "that he carries the stars within" remains the subject of the stars above, thus unable to fulfill the high destiny set for him by God (Paracelsus 1951, 128, 129, 139, 154).

Paracelsus, like many early modern thinkers, was an ardent millenarian—that is, he believed on biblical authority that God would sooner or later bring in the millennium, a thousand-year reign of peace, justice, and prosperity. Paracelsus believed that this great change would come fifty-eight years after his death and that humans, as seekers of natural knowledge, would contribute to making it happen, a process that he thought of in alchemical terms: "Man must bring everything to perfection. This work . . . is called 'alchemy,'" that is, the discovery of chemical secrets and, ultimately, the legendary philosopher's stone, to be tapped for the good of all. In medical terms, this process should lead to unheard of cures and the prolonging of life; in material terms, to unheard of plenty and a moral economy in which everyone's "natural needs" would be met. This millennial theme would be closely associated with the development of science throughout the early modern period, especially in Protestant circles (Trevor-Roper 1985, 157; Paracelsus 1951, 92–3, 168, 176–7, 218).

Paracelsianism spread with the Protestant Reformation. Paracelsus himself did not identify with any church. He pursued instead a nonsectarian, ecumenical ideal that would unify all people in preparation for the alchemical millennium. But his attack on ancient authority and the medical establishment, his use of the vernacular, and the trust he put in ordinary people, their practical knowledge and the dignity of their labor, had their counterparts in the Reformation movement. After all, Martin Luther himself (1483–1546) had said that "any potter has more knowledge of nature than these books [of Aristotle]" (Trevor-Roper 1985, 165–6; Copenhaver and Schmitt 1992, 39).

The resonance between Paracelsus's ideas and the Radical Reformation was particularly strong. The radical reformers were self-professed saints, often known as Anabaptists, who were convinced of their salvation and claimed that, as God's chosen ones, they were illumined and empowered by God or Holy Scripture without need of earthly interpreters. On that basis, they sometimes went on to separate from the state church, to organize into congregations of like-minded believers, and to attack established authority in the name of social justice and equality. Thus, German peasants, inspired by such radical thinking, rose up against rural priests and landlords in the great Peasants' Revolt of 1525.

Paracelsus's writings chimed with this radical thinking. (He also sided with the peasants against their oppressors.) He said, for instance, that true philosophy comes "from Holy Scripture" and not from Galen and that God gives virtuous, industrious people the power to "perform miracles on earth . . . for man is of divine nature" (Trevor-Roper 1985, 150; Paracelsus 1951, 195–6). The clerical and medical elites, on the other hand, are vain and greedy, bankrupt in head and heart. They exploit the poor by pretending to knowledge they do not have. The common people are spiritually and intellectually superior to their social betters. If the notables would reform themselves, they would do well to go to peasants and artisans to see faith and morals in action, true charity demonstrated, and to imbibe a genuine knowledge of nature. Following Paracelsus, radical healers often backed their claims to knowledge and authority by denouncing the pagan authority of Galen and Aristotle and by appealing instead to a Bible-based science and their own inspiration. Unlicensed medical practitioners also found support, in Paracelsus's views, for their attack on the medical monopoly established by university-trained physicians (Trevor-Roper 1985; Webster 1982, 48–58; 1993; Williams 1992).

The Copernican Revolution

Paracelsus and his followers were not the only rebels. A quieter, but no less potent, rebellion in astronomy began in 1543 with the publication of *On the Revolutions of the Heavenly Spheres* by the Polish priest Nicolaus Copernicus (1473–1543). Upon reading the book, Luther denounced its author as "an ass" intent on turning the world upside down (Wightman 1972, 117). Reviving Aristarchus's ideas, Copernicus propounded a picture of the universe with a central stationary sun. Deeply respectful of Ptolemy, he wished to do no more than simplify and purify the Ptolemaic scheme while preserving the spherical framework and uniform circular motion of the heavens. To do so, Copernicus made three major changes. First, he put the earth in motion around the sun. Second, he also gave the earth a daily rotation on its axis. Third, he eliminated the equants that Ptolemy had introduced into the system, which Copernicus regarded as disfigurements (Cohen 1960, 36–63).

But his reform went only so far. For instance, to explain the motion of the planets, he retained the planetary spheres and epicycles of the Aristotelian-Ptolemaic model, and, having eliminated the equants, he was forced to add more epicycles to account for some of the apparent motions of the planets. The Copernican picture was a halfway house in yet another respect. According to ancient and medieval tradition, the heavenly spheres move by virtue of their natural perfection. Breaking with this tradition, Copernicus argued that the sphere of the fixed stars does not move at all. This seemed illogical (how can something whose nature it is to move, not do so?). But he kept the starry sphere anyway and maintained that it serves as the envelope that holds the world together. It also allowed him to say that the universe, though immeasurable, is not infinite but bounded (Koyré 1968, 28–35).

During the last quarter of the sixteenth century, others took the step that Copernicus refused to make and asserted that the sphere of the fixed stars of Copernican astronomy does not exist and that the starry heavens are, in fact, infinite. The first to do so was the English astronomer Thomas Digges (ca. 1543–95), who, in 1576, published a book in which he said that the starry heavens extend to infinity. He went on to identify this astronomical heaven with the habitation of God, the angels, and saints in paradise. His was not a conception of a natural infinite but the conflation of a natural and supernatural one (Fig. 7) (Koyré 1968, 35–9).

The Copernican system did have at least two major advantages over

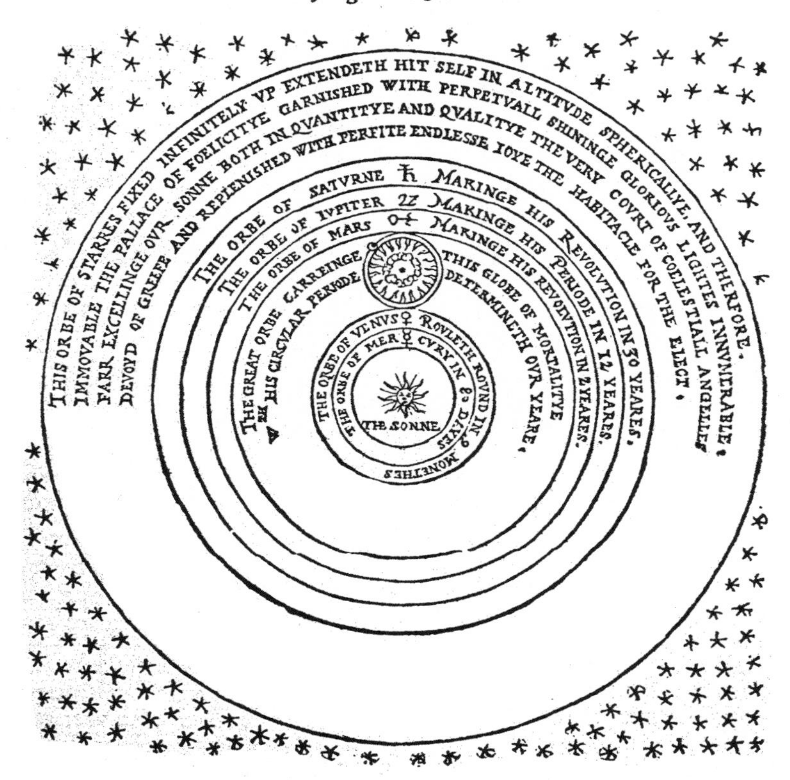

FIGURE 7 Digges's Picture of an Infinite "Copernican" Universe. The starry sphere is infinitely extended and is identified with "the habitacle for the elect that is, the Christian heaven or paradise." (From Digges's *A Perfit Description of the Caelestiall Orbes,* 1576.)

the Ptolemaic system. First, the diurnal rotation of the earth explained the apparent daily motions of the stars and planets. Second, the apparent retrograde motion of the planets in their course around the sun—the way they sometimes reverse direction as seen against the background of stars—could now be accounted for. According to Copernicus, the earth is also orbiting the sun, and each planet has a longer or shorter period of revolution that varies directly with its distance from the center. So, when the earth passes a planet, say Mars, which is further from the sun, that planet appears, to a terrestrial observer, to reverse direction.

These advantages, however, could scarcely be said to outweigh the objections one could raise as to the validity of the Copernican model, objections similar to the ones Aristarchus had faced so many centuries before. There was again no observed parallax of the fixed stars (daily or seasonal change of their position relative to the earth), which meant either that they were exceedingly distant or that the Copernican hypothesis itself was a fallacy.

More fundamental, Copernicus had thrown a wrench into the works by introducing a moving earth into an Aristotelian universe. He had to say that the earth is just another planet. But to say that was to deny the binary universe, that is, to deny that the earth and the heavens are made of different elements, are subject to different sets of physical laws, and behave, therefore, in different ways (Cohen 1960, 60). How could the earth, as conceived by Aristotle, move in circles? To do so was to defy its nature as a heavy, immovable thing. And, if the moon circled the earth, as Copernicus agreed, what kept it in orbit while the earth plunged through space in its race around the sun? Subscribing to Renaissance magic, Copernicus endowed the sun with a mysterious governing power. But his heliostatic hypothesis did not square with the Aristotelian framework to which it was still attached. His new world-picture may be said to have raised more fundamental questions than it answered (Cohen 1960, 36–63; Kuhn 1957).

The reaction to his book, in the three decades following publication, was subdued, in part because it was prefaced by an anonymous foreword (which Copernicus did not write) claiming that the book was to be regarded as a mathematical exercise showing how planetary motions might be explained rather than as a true picture of the universe. But, in time, others, like Digges, seized on the possibilities for new thinking about the universe opened up by Copernicus; the boldest of these was Giordano Bruno (1548–1600).

Bruno's Infinite Universe

By 1584, Bruno had become an ardent champion of his own striking brand of Copernicanism. Drawing on Lucretius, Bruno revived the ancient Epicurean idea of the infinity of the universe. But Bruno's system, though spatially infinite and strewn with innumerable worlds made of an infinity of atoms, was otherwise quite different from the Epicurean picture. The latter was composed of lifeless particles moving randomly through vast spaces inhabited by gods indifferent to the world. But Bruno, echoing Nicholas of Cusa (1401–64), built his universe around the principle of plenitude. An infinite and omnipotent cause, Bruno's God could do no other than express Himself in an infinite universe made up of countless worlds, not dead, but like the earth, teeming with all possible forms of life (Koyré 1968, 6–23, 40–3, 54; Schmitt and Skinner, 1988, 548–56).

Our solar system, according to Bruno, is one such world, with the earth circling the sun, as Copernicus said. Aristotle was dreadfully wrong; the universe is not finite and immutable with the earth immobilized at the center. On the contrary, Bruno proclaimed, "there are no ends, boundaries, limits or walls which can defraud or deprive us of the infinite multitude of things." The universe is constantly undergoing change, which provides the fuel for cosmic "renewal and restoration." Infinity is the ultimate guarantee, "for from infinity is born an ever fresh abundance of matter" (quoted in Koyré 1968, 44). Nor do things die; they merely dissolve into and reemerge from the cosmic soup, "for the material and substance of things is incorruptible and must in all its parts pass through all forms" (Yates 1964, 242). Thus, Bruno spurned the Christian doctrine of a supernatural heaven and hell and revived the Pythagorean idea of metempsychosis.

Bruno's chief target was Christian theology. God did not create the world; indeed He could not because, though He is transcendent, He is also immanent in nature, which is thus not his creature but his self-manifestation. The atomic particles that fill the universe are alive—not only animate but laced with divinity. The world is made of infinite, divinized atoms moving in infinite space and forming infinite worlds pullulating with life. All space, moreover, is not empty but full, a plenum of material ether, binding together the elements of the universe (Grant 1981, 189).

To orthodox Christians, Bruno's Copernicanism was heretical. The earth revolves around the sun, which is at the center of our solar system but not of the universe itself because infinity has no center. Worse, the celestial bodies that make up the innumerable worlds are self-moving, not, that is, attached to tracks or spheres, as Copernicus claimed, but whirling through space under their own steam, propelled by inherent, occult power. All nature is self-activating and self-regulating because it is divine.

Bruno was a Dominican friar, but his philosophy was pagan, based on a mixture of Lucretius, Pythagoras, and the revival of Hermetic philosophy begun by the Florentine Neoplatonists Ficino and Pico. Where they had tried to harmonize Hermetism and Christianity, Bruno discarded the Christian elements in favor not of the watered-down Ficinian magic but of the full-blown pagan magic of the supposed Egyptian priest Hermes Trismegistus. Here was the wisdom from which Bruno derived his pantheistic, or vitalistic, materialism and his belief, now reinforced by Copernicus, that everything in nature moves, including the earth. For Bruno, this Egyptian wisdom was not just a matter of private belief, a secret heresy, so to speak. In his travels throughout western Europe, his visits to princely courts and universities, and his books, he never ceased to expound and promote his pagan revelation. Railing against both Catholic and Protestant Christianity, he made enemies wherever he went because of his virulent impiety. He was finally arrested, tried, and convicted for heresy by the papal Inquisition—and burned alive at the stake in 1600.

Like Paracelsus, Bruno was both a natural philosopher and a moral reformer, and his social ethics deserve brief examination. Perhaps following Paracelsus, and certainly echoing him, Bruno took up the theme of the Fortunate Fall. But where Paracelsus interpreted the idea in entirely Christian terms, Bruno conflated the biblical account with the pagan version provided by Virgil's *Georgics* (29 B.C.E.). There, the god Jupiter brings to an end humankind's life of ease in the Golden Age so that afterward people are driven by necessity to work and to use their wits in order to survive. But, out of these conditions of hardship, their minds are sharpened, industries invented, skills discovered. In this process, there is the potential for both material and moral improvement. The former is sooner achieved than the latter, but, if the right steps are taken, moral reformation may follow (Bruno 1992, 205–6; Low 1985, 137–8; Virgil 1982, 52–3).

Looking out on late sixteenth-century western Europe, Bruno diagnosed its moral ills and prescribed some drastic cures. Most fundamental, there was the problem of Protestantism, especially the version produced by John Calvin (1509–64), with its key doctrine of predestination, the belief that not even good works can save those destined for hell. According to Calvin, human beings, tainted by original sin, are entirely without merit in the eyes of God and can only be saved by divine grace. Here was an idea, Bruno thought, that sold human beings short by denying their true nature. After all, "the gods had given intellect and hands to man and . . . made him similar to them." Thus, instead of wallowing in a Calvinist-induced despair, people must set about using these divine gifts of head and hand to understand and control nature for human purposes. To the extent that this is done, man can "succeed in preserving himself as god of the earth" (Bruno 1992, 205).

Bruno echoed the optimistic anthropocentrism of the Florentine Neoplatonists and added a new historical dimension to their outlook. The world was better off before the Reformation; the old Catholic theology of good works, while falsely based on "the pedantry" of Aristotle, was still preferable to the Protestant emphasis on predestination, the doctrine that human efforts avail nothing and that God has already decided who will be saved or damned. To overcome that doctrine's destructive effects, what must be rekindled in human hearts is the ancient "appetite for glory," the very appetite starved by the Reformation, "for this . . . is the only . . . spur that is wont to incite men and fire their enthusiasm for those heroic deeds which enlarge, maintain, and fortify republics" (Bruno 1992, 124–5, 146; Yates 1964, 226–7; 1982).

Bruno also addressed the problem of what constitutes the good life. His political ideal was embodied in the ancient Roman Republic. But, failing this, he urged that, whatever the precise form of government, it be one in which "the weak be not oppressed by the stronger; . . . the poor be aided by the rich; . . . virtues and studies, useful and necessary to the commonwealth, be promoted . . . ; and . . . the indolent, the avaricious, and the owners of property be scorned and held in contempt." Bruno's views on wealth and poverty remind us of Paracelsus, and, in fact, he saw Paracelsus as, to some degree, his precursor (Bruno 1992, 144–5; Yates 1964, 251).

The Origins of Modern Skepticism

During the late sixteenth century, even as Bruno was spreading his message across Europe, an intellectual movement was developing that would deal a serious blow not only to Bruno's philosophy but to much else. This new current was a revival of an ancient skepticism that claimed that nothing can be known with certainty. The roots of this movement ran deep. The new thinking about nature, like Bruno's, was partly responsible for it. If the universe was said to be infinite, where did this leave the closed world of Aristotle that had been taught for centuries? Who was right about the relative places of the earth and the sun, Ptolemy or Copernicus? And which medical regimen would keep you alive longer, that of Galen or Paracelsus?

The Skeptical Crisis

There were at least three other developments that underwrote the skeptics' argument: the Reformation, the overseas discoveries (especially in the New World), and the recent publication of certain ancient texts of skeptical philosophy. First, what was the effect of the Reformation? Catholics and Protestants alike looked to the Bible as the word of God and therefore the final authority in matters of faith. But both acknowledged that the Bible had to be properly interpreted in order for its truth to be revealed, and their approaches to this enterprise were fundamentally different. For centuries, the Catholics had held that the Bible must be read in the light of papal pronouncements and traditions established by church councils. But Protestants, beginning with Luther, now claimed that what conscience is compelled to believe on reading the Bible is true (Popkin 1979, 3). Catholics quickly pointed out that if the Protestants had their way and individual conscience became the measure of truth, every man (or woman) could become his (or her) own church, and the result would be anarchy. Events, moreover, seemed to bear this argument out.

By the late sixteenth century, many parts of western Europe had paid a heavy price for religious dogmatism. Violence and terror had taken two forms—religious wars, which engulfed whole territories, France and the Netherlands in particular, and the more limited madness known as the witch-craze, the trial, torture, and execution of those accused of witchcraft, 80 percent of whom were women. The misery of the wars and the witch-craze produced, in some minds, a revulsion against dogmatism and a growing skepticism about where truth could be found.

The second factor, working in the same direction, was the overseas voyages, which caused thinkers to see that truth is relative to culture, that what one people takes for good, beautiful, and true may be thought of as the reverse by another. Third, the newly available writings of Sextus Empiricus (fl. 200 C.E.) had their influence on the origins of modern skepticism. Sextus's writings are the sole surviving texts of a movement named for its supposed founder, Pyrrho of Elis (ca. 360–275 B.C.E.). The Pyrrhonists held that human beings should suspend judgment on any question to which there is no clear answer, including the question whether anything can be known at all. Thus, they saw themselves as answering both those who were confident that something can be known and those who were equally confident that nothing is certain. For the Pyrrhonist, this suspension of judgment is meant to lead to a psychological stance of *ataraxia,* freedom from anxiety. The Pyrrhonist escapes dogmatism by following "his natural inclinations, the appearances he is aware of, and the laws and customs of his society" (Popkin 1979, xv). Such an outlook was intended to breed detachment, privacy, and quietism. Sextus's writings were almost unknown in the Middle Ages, but Greek manuscript versions entered Italy in the fifteenth century, and a Latin edition of his *Hypotyposes* was published in 1562, followed by a Latin edition of all his works in 1569 (Popkin 1979, 1–41).

Montaigne's Response

Eleven years later, the French thinker Michel de Montaigne (1533–92) published the first two books of his *Essays,* which were to have a major impact on religion and natural philosophy. Montaigne, who retired to his study in 1571, looked out on his world and despaired. He had experienced the horror of the French civil wars, which broke out in 1562, at close range in his native Bordeaux; he also reacted negatively to the witch-craze: "It is," he said, "taking one's conjectures rather seriously to roast someone alive for them" (quoted in Burke 1981, 24). And he had read the skeptics—Cicero, Desiderius Erasmus (1466–1536), Henricus Cornelius Agrippa (1486–1535), and, most recently, Sextus himself. The combination of tragic events and philosophical reading left Montaigne reeling. Nothing in his thought or experience, he felt, could be taken as immutably certain. Laws and customs are diverse and contradictory and vary with place and time. The religious wars showed that what is taken for divine truth one year may be condemned as execrable heresy the next. He said further, "What am I to make of a virtue that . . . becomes a crime on the other side of the river?" (quoted in Schneewind 1990, 1:43).

In Montaigne's view, the human faculties—reason, the senses, and passions—are all equally unreliable avenues to knowledge and truth. Reason "is like a tool" that can be adapted to support any clever argument or theory. For example, "where the first principles of Nature are concerned, I cannot see why I should not accept, as soon as the opinions of Aristotle, the 'Ideas' of Plato, the atoms of Epicurus, the plenum and vacuum of Leucippus and Democritus, the water of Thales, the infinity of Anaximander, . . . the numbers . . . of Pythagoras . . . or any other opinion drawn from the boundless confusion of . . . doctrines produced by our fine human reason" (Montaigne 1993a, 114, 144). The result is a scientific free-for-all, with truth the loser.

The senses, like reason, are totally untrustworthy. First, how do we know that we have them all? If we do not, there is no way of telling what we are missing. How, for example, can a blind man know what sight teaches? Second, we find that even the ones we do have can often be misleading under normal conditions and, even more so, when affected by the passions or poor health. Finally, Montaigne makes an argument that will be basic to the Scientific Revolution. A sense impression and the outside object that produces it are two quite different things. "So whoever judges from appearances judges from something quite different from the object itself" (Montaigne 1993a, 186). The only things we have direct knowledge of are our sensations, and not the world they seem to represent.

If reason and the senses cannot tell us about nature, neither reason nor the passions can show us how to live. Following the ancient Epicureans, Montaigne claimed that "nature needs wonderfully little to be satisfied." But human beings, left to their own devices, are insatiable: "The lawless flood of our greed outstrips everything we invent to try and slake it." For Montaigne, unlike Bruno and Paracelsus, the Fall from paradise was an unmitigated disaster, the result of our excesses and of the worst of our passions, which is pride. Montaigne was taking aim at the Renaissance argument for the dignity of human beings. To his mind, man is "this miserable and wretched creature, who is not even master of himself . . . and yet dares to call himself lord and emperor of a universe, the smallest particle of which he has no means of knowing, let alone of swaying." Arrogance and vanity are the rule, and "in nothing does man know how to stop at the limit of his need," including "the curiosity for knowledge," whose excessiveness caused the Fall (Montaigne 1993a, 13, 23, 36, 64; Schneewind 1990, 1:58).

Here is our predicament. Our natural faculties yield nothing but uncertainty, but our arrogance mistakes for knowledge the misinformation we are fed by our senses and reason. At this point, our greed takes over and causes us to act on this misinformation to pursue "imaginary and fantastical" goods that exceed our natural limits and fail to satisfy our natural needs. "False opinions and ignorance . . . have poured so many strange desires into us that they have chased away almost all the natural ones" (Montaigne 1993a, 37, 50).

But there is a way out of this trap. Montaigne advises that we do as he does and follow the Pyrrhonist plan he has read about in Sextus, which amounts to "doubt and suspense of judgment" on everything.

> This leads to . . . *ataraxia*: that is a calm, stable rule of life, free from all the disturbances (caused by . . . such knowledge as we think we have) which give birth to fear, avarice, envy, immoderate desires, ambition, pride, superstition, love of novelty, rebellion, disobedience, obstinacy and the greater part of our bodily ills. (1993a, 70)

Pyrrho's prescription does not create a wooden monster, a benumbed automaton, such as one might suppose, "but a living, arguing, thinking man enjoying natural pleasures . . . of every sort and making full use of all his parts, bodily as well as spiritual—in, of course, a right and proper way" (Montaigne 1993a, 73).

This Pyrrhonian "way" should also induce us to obey established authority in both church and state. As Montaigne says, "I . . . remain where God put me. Otherwise I would not know how to save myself from endlessly rolling." Given man's weaknesses, "duty must be laid down for him, not chosen by him." Montaigne is a religious fideist; he accepts the authority of Catholic Church not by reason but on faith: "Only humility and submissiveness can produce a good man." Montaigne makes it clear that he is following the entire Pyrrhonian regimen—accepting authority from above, taming passion, and pricking pretension (Montaigne 1993a, 53, 149; 1993b, 377).

But there is more to obedience than mere submission to God's will and to political authority. Montaigne willingly limited his freedom in the public space so that he could carve out for himself a private space in which he could live and think undisturbed by the outside world: "The wise man ought to retire into himself, and allow himself to judge freely of everything, but outwardly he ought completely to follow the established order" (quoted in Burke 1981, 27). The trade-off is clear—intellectual autonomy in return for outward conformity, the opportunity to cultivate the self at the price of abandoning public life. Montaigne was no civic humanist, like Erasmus; still less a radical activist, like Bruno or Paracelsus. As we shall see in chapter 4, his skepticism had revolutionary intellectual consequences, while scarcely producing so much as a social ripple.

Montaigne was influenced by the ancient Epicureans, especially Lucretius, who retreated from the world, and by the ancient Stoics, who armed themselves against it. In fact, like Seneca (one of Montaigne's favorite sages), he saw no essential incompatibility between the ethical teachings of these two classical schools (Allen 1944; Schmitt and Skinner 1988, 367, 375, 381, 774). Montaigne was also influenced by the destructive forces unleashed by the religious wars and the witch-craze and by what he regarded as the extravagant claims to occult wisdom made by seers like Ficino, Pico, Bruno, and Paracelsus. Against all such threats to sanity, Montaigne lowered the stakes and quoted St. Paul (Romans 12:3): "Be not wiser than is becoming, but be soberly wise." Given the environment created by the Reformation, here was a formula with a big future for the next hundred years (Popkin 1979, 42–65; Screech 1991).

Charron's Response

Pierre Charron, in *Of Wisdom* (1601), adopted Montaigne's skepticism and took it in some striking new directions. Montaigne's motto was "*Que sais-je?*" ("What do I know?") But Charron said: "*Je ne sais.*" ("I do not know.") He was even more emphatic than his master that human beings can know nothing with certainty. He followed Montaigne's arguments to back up the point and extended one, the argument from cultural relativism, much further than Montaigne was willing to take it: "The great variety, diversity, and inequality amongst men . . . and the vast range of laws and customs that have established themselves in the world" means that "there is no better school of life than to observe . . . the diversity . . . and to relish the perpetual variety to which our nature lends itself" (quoted in Gregory 1992, 91). Charron's idea of human "diversity" relativizes culture, just as Bruno's "plurality of worlds" relativizes nature, and the result, in both cases, is heretical.

For Charron, religions are especially important pieces of this baggage of custom, because they have such great authority among the common people. Christianity, like every other creed, is relative to its culture. Montaigne accepted Catholicism on faith as the one true, revealed religion. But Charron did not and could not follow him in making that leap of faith because, for him, unlike Montaigne, all religions come close to being nothing more than human inventions, cultural artifacts, "maintained by human means and preserved by human hands." This does not mean that much craft or care goes into their construction: "All religions have this in common, that they are an outrage to common sense, for they are pieced together out of a variety of elements, some of which seem so unworthy, sordid, and at odds with man's reason that any strong and vigorous intelligence laughs at them" (quoted in Gregory 1992, 97, 99, 100). The most effective religions are those that are constructed to preserve the social order.

Religions, however, do have a function. They are useful for controlling the masses who fall for them because they are "slaves to custom." But a small elite of wise men can see the situation for what it is, conform outwardly to the public creed, and "behind this mask . . . continue the free exercise of reason, making responsible choices in the light of conscience." This is not, however, the conscience of the Protestants, the spiritual principle accessible to each individual. Instead, Charron's conscience is the ethical principle of secular reason, accessible only to the enlightened and tough-minded few. With Charron, morals (at least for the few) have been cut loose from religion: "I want men to be good and to be . . . firmly dedicated to what is right . . . , motivated by love of themselves and by the desire to live up to their nature as men" (quoted in Gregory 1992, 101, 107). Here was the definition of a secularized conscience as dangerous as the sacred version so dear to Protestantism. Charron's views were recognized for what they were by his Catholic enemies, and his "scandalous book" was placed on the *Index* in 1605; Montaigne's *Essays* had to wait until 1640 before receiving the same treatment. As we shall see in chapter 4, the greatest natural philosophers of seventeenth-century France would develop their views in response to the skepticism of Montaigne and Charron.

The New Science

New Astronomy

REVOLUTIONARY CHANGES OCCURRED in many fields of science in the late sixteenth and early seventeenth centuries, and nowhere more than in astronomy. Copernicus and Bruno, as we saw, were among the first to blaze new trails in the study of the heavens. They were followed by Tycho Brahe (1546–1601), Johannes Kepler (1571–1630), and Galileo Galilei (1564–1642), who are the subjects of the first section of this chapter.

Brahe

Astronomers observed a bright new star in 1572, and five years later they tracked the path of a new comet. In both cases, careful observation made clear what was, according to Aristotelian physics, not supposed to happen. A new star defied the Aristotelian doctrine of the changeless heavens, and comets had been observed before, of course, but were always ascribed to the sublunary sphere where the change they represented could be accommodated. The Danish astronomer Tycho Brahe was equipped with the best observatory in Europe, Uraniborg, funded by the king of Denmark. He carefully calculated the distances of the star and comet from the earth and judged that both were superlunary phenomena. Change was occurring in the "changeless" heavens. The case of the comet was especially disturbing to the old thinking: Astronomers watched as it swept right through what were supposed to be solid ethereal spheres, the backbone of the Aristotelian universe. As a result, Tycho declared that those spheres do not exist.

He also devised his own new picture of the heavens. Still unable to accept a mobile earth, he depicted a cosmos in which the sun and moon circle a stationary earth, but one in which all the other planets revolve around the sun (Thoren 1990).

If Bruno were right, the cosmos is not binary but uniform throughout. If the heavenly spheres do not exist, moreover, there were two other big questions to be answered. First, what exactly is the nature of the planetary orbits, and, second, what holds the planets in those orbits? To put it another way, why do the planets stay in orbit, whether around the earth or the sun, rather than either falling into the center or flying off into space?

Kepler

Kepler eventually became a devoted Copernican who dedicated himself to observing the planetary orbits and to answering the new crucial questions about them. But he had not always been a convinced Copernican. In 1600, Kepler had become Tycho Brahe's research assistant, and, when the latter died, he inherited Tycho's rich trove of research data on planetary orbits, especially that of Mars. At first, Kepler tried to reconcile these data with the Tychonic picture of the heavens. But, failing to make the fit, he rejected his master's scheme in favor of his own version of Copernicanism. Influenced by Renaissance Neoplatonism, Kepler, like Copernicus, attributed to the sun a mysterious, godlike power that makes it the logical center of the cosmos. Like Copernicus, Kepler, too, held that the universe, though very large, is finite. But, unlike Copernicus (and Galileo), Kepler ultimately rejected a universe based on the principle of circularity.

In 1609, he published his greatest work, *Astronomia Nova* (*New Astronomy*), in which he discarded the perfect spheres, circular orbits, and uniform speeds of Copernicus and the ancients. Kepler had discovered the first of his three laws of planetary motion; a combination of careful observation and mathematical calculation showed that the orbits of the planets can best be understood not as complex combinations of circles but as simply elliptical. An elliptical orbit means that a planet's distance from the sun constantly changes, and Kepler discovered that there is a connection between this distance and orbital velocity. The planets speed up as they approach the sun and slow down as they recede from it, and do so in a mathematically regular and predictable way. Hence Kepler's second law, also announced in his

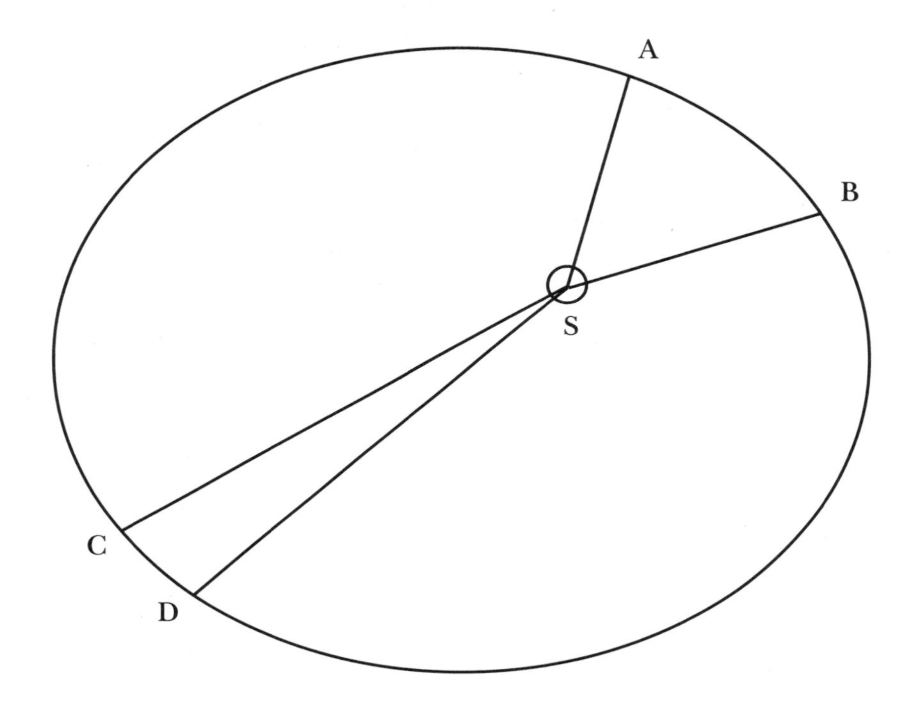

FIGURE 8 Kepler's First Two Laws of Planetary Motion. First, a planet revolving around the sun at one focus describes an elliptical, and not a circular, orbit. Second, in the course of its revolution, it sweeps out equal areas in equal times such that ASB is equal to CSD, which means that the planet speeds up as it draws closer to the sun and slows down as it recedes from the sun—and does so in a mathematically regular way.

New Astronomy: The planets, moving in their elliptical orbits around the sun, sweep out equal areas in equal times (Fig. 8). Kepler stated his third law ten years later in *Harmonice Mundi* (*Harmonics of the World*). There he held that the square of the time of revolution of a planet around the sun is proportional to the cube of its mean distance from the sun. For Kepler, his laws, especially the third, demonstrated his Christianized, Pythagorean-Neoplatonic belief that God created the universe to conform to principles of order and harmony that may be expressed mathematically. "I feel carried away," he said, "and possessed by an unutterable rapture over the divine spectacle of the heavenly harmony" (quoted in Caspar 1959, 267).

But, for Kepler, these laws, despite their mathematical elegance and explanatory power, represented a kind of fallback position. They explained *how* the planets move, but he had set out to discover the force that causes them to move and to establish, mathematically, "one universal force law" to explain *why* they move as they do (Holton 1973, 72). Had he succeeded, he would, in effect, have replaced the old binary picture with a completely unified system of the world based on mathematical calculation. Kepler's aims were anything but modest, and he failed after enormous effort to reach them. It was Newton, following in Kepler's footsteps, who finally succeeded in 1687. But it was Kepler who first conceptualized the problem and tried to solve it (Holton 1973).

Kepler was influenced by the English philosopher William Gilbert (1540–1603) and his great book *De Magnete* (*On Magnets*), published in 1600—the first important work on experimental science published in England. Gilbert demonstrated the nature of magnetism in some fifty experiments. He conceived of the earth as a gigantic magnet. Gilbert was not a Copernican but accepted the diurnal rotation of the earth, which he thought was caused by the earth's own magnetic energy. According to Gilbert, echoing the Hermetic tradition, the universe and everything within it is self-animated, and magnetism is a function of the earth's self-animating soul.

Kepler extended Gilbert's theory in a new direction. If the earth can display such magnetic power, Kepler thought, then so can the sun. The force keeping the planets in orbit might be the sun's magnetism. After all, one magnet attracts another even when separated by some distance; the closer they are, the stronger the pull, and the farther away, the weaker the pull. Here was a model for understanding the planetary orbits around the sun and the changes in velocity as the planets describe their elliptical orbits. Kepler imagined that this magnetic power radiates out from the sun not in all directions, as solar light and heat do, but only along the plane of the ecliptic, the plane in which all the planetary orbits lie.

Kepler's achievement had little immediate positive impact. His great contemporary Galileo, as we shall see, spurned his conclusions. But he had tried seriously to understand the behavior of the planets in a heliostatic universe (the sun is slightly off center), now shorn of the spheres that had once kept the planets on track. He had also broken the hold of ancient astronomy, more decisively than Copernicus, by overturning the hallowed principle of circularity and replacing it with elliptical orbits whose behavior could be accounted for mathematically. Kepler's laws represent a triumph for mathematical description. Galileo's science would further demonstrate the power and utility of numbers (Holton 1973; Stephenson 1994).

Galileo

Telescopic Discoveries

Galileo had become an avowed Copernican by perhaps 1597 and spent much of the rest of his career championing the heliostatic universe (but not the Keplerian version). During the first decade of the seventeenth century, while professor of mathematics at the University of Padua, he made close and systematic observations of the heavens and discovered another new star like the one seen in 1572. As soon as he learned of the invention of the telescope in 1608, perhaps by Hans Lippershey (ca. 1570–ca. 1619), a Dutch lens grinder, he devised one for himself. Galileo spent the next year observing the heavenly bodies and published his findings in *Sidereus Nuncius* (*The Starry Messenger,* 1610), written in the Italian vernacular for all to read. It became one of the defining books of the Scientific Revolution because of its polished style and startling contents, which, in a number of ways, contradicted the old astronomy and provided fresh evidence for (but did not prove) the mobility of the earth in a heliostatic system (Van Helden 1977).

Through Galileo's telescope were revealed not the perfect, ethereal globes of the ancients but the rough surface of the moon, which looked remarkably like the earth. To the same effect, Galileo reported, in 1613, that there were spots on the sun, and the careful observation of these spots proved that the sun rotates on its axis. The old binary world-picture was becoming less credible with each year. The telescope also showed that the dark side of the moon (facing away from the sun) reflected earthshine (solar light bouncing off the surface of the earth). In 1610, the question of whether the planets have their own light, like the sun and stars, or reflect light, like the moon, had not yet been decided, and Galileo's discovery of earthshine now suggested that the latter was true (Cohen 1960, 67–76).

Perhaps his most notable telescopic finding was his discovery of four moons circling the planet Jupiter. Knowing which side his bread was buttered on, he called them "the Medicean stars" after the family of his patron Cosimo II de' Medici, the grand duke of Tuscany (1609–21). The name was especially well chosen in view of the fact that, since its founding, the Medici dynasty had associated itself, for reasons of self-promotion, with the planet Jupiter and its astrological virtues. So Galileo, in discovering its moons, was seen, and wished to be seen, as confirming the family's cosmic status by revealing what was already written in the heavens. And he was soon rewarded. By July 1610, a few months after his book was published, Galileo had been appointed philosopher and mathematician to Cosimo II and was earning one of the ten highest salaries at the ducal court. The title of court "philosopher" was particularly important to Galileo because, in the Scholastic scheme of things, philosophy was accorded higher status than mathematics, with its practical associations.

But there was much more at stake than Galileo's personal advancement, and he knew it. In naming the moons of Jupiter and receiving his reward, "he turned Medici power to the legitimation of his discoveries and his telescope" (Biagioli 1993, 127). Book publishing was fine, as far as it went, for getting out the truth. But, in a court-centered aristocratic society, like that of Florence, acceptance of scientific novelty came on the wings of patronage, and there was no better source of it than the man at the top of the social heap. What was seen through a newfangled telescope was more likely to be credited once the grand duke had given it the nod. Galileo also knew how to exploit his new court connections; he had telescopes and copies of his book sent to European rulers through Medici diplomatic channels. In getting his message across, he worked from the top down (Biagioli 1993, 1–157).

The discovery of Jupiter's moons struck another blow against geocentrism: The earth lost its unique status as the central body around which all others revolved. The Medicean stars could also be used to support the argument for a moving earth. As the historian I. B. Cohen explains, "It was . . . plain that in every system of the world that had ever been conceived Jupiter was considered to move in an orbit, and if it could do so and not lose four of its moons, why could not the earth move without losing a single moon?" (1960, 80).

Galileo's telescope also revealed the phases of the planet Venus. In the Ptolemaic system, Venus would always be located between the earth and the sun's sphere and would always appear to be the same size when observed with the naked eye. But, through the telescope, Venus appears differently at different times of the year, sometimes as a small disk, sometimes as a large, faint crescent, and sometimes as something in between as to both size and shape. These changes, or phases, can be explained if we assume that Venus revolves around the sun and not the earth. For instance, when Venus is observed as a small, bright disk, we may assume that it is on the far side of the sun from the earth and reflecting the sun's light, and when it is observed as a much larger crescent, we may assume that it is now between the earth and the sun and lit mainly on its far side. These phases could never be accounted for by the Ptolemaic system but were entirely consistent with a Copernican one (Cohen 1960, 81–3).

Physics

Galileo's discoveries represented a stunning performance and won him considerable fame. But, for him, they also created a new challenge. Now that there was so much new and powerful evidence for Copernicanism, the next question was, What were the physics of a moving earth? On a common-sense level, how can the earth be in such rapid motion, spinning on its axis and hurtling through space, when it does not appear to be moving at all? On an intellectual plane, what must take the place of the physics of the Aristotelian binary universe if the earth can no longer be thought of as uniquely at rest nor the heavens immutably perfect, each governed by a separate set of laws? Here was a challenge that Galileo was to pursue in two more great books, *Dialogue Concerning the Two Chief World Systems* (1632) and *Discourses on Two New Sciences* (1638).

Galileo never explained why the earth can move, but he was successful in showing why terrestrial experiments such as the dropping of weights from high places can neither prove nor disprove the motion of the earth. The old objection to a moving earth had been that, if a ball were dropped from a tower on a moving earth, it would not land at the foot of the tower but well behind. In fact, we know that the ball falls straight down to the base of the tower. To answer the problem, Galileo took the case of an object dropped from the top of a ship's mast. Whether the ship is at rest or in constantly uniform motion, the object will land at the base of the mast. In the case of a ship moving at a uniform speed, the object dropped from the top of the mast participates in the ship's motion and thus travels forward at the exact speed of the ship. Exactly the same goes for the ball dropped from a tower. "Keeping up with the earth," Galileo said, "is the primordial and eternal motion ineradicably and inseparably participated in by this ball as a terrestrial object, which it . . . will possess forever" (quoted in Westfall 1977, 17). Galileo stated elsewhere that he had performed experiments duplicating the case of dropping an object from a ship's mast. But, in the *Dialogue*, he simply declared: "I, without observation, know that the result must be as I say, because it is necessary" (quoted in Cohen 1960, 94).

For him, experiments functioned as confirmations of what could be arrived at by reason, especially when reason was backed up by mathematics. It was by such mathematical reasoning that Galileo discovered his law of uniformly accelerated motion. All bodies, he found, regardless of differences of weight, fall at the same rate of acceleration under ideal conditions (that is, in a vacuum where resistance has been eliminated). Galileo thus dealt a decisive blow to Aristotelian physics, which held that the rate of fall is a function of weight (heavier bodies fall faster than lighter ones). Equally important, he discovered that all falling bodies obey a mathematical law of uniform acceleration: The distances traversed in intervals of time by a body falling from rest with a uniformly accelerated motion are to each other as the squares of the time intervals. In fact, such acceleration occurs at the rate of thirty-two feet per second of velocity gained each second. Only after working out the mathematics of the acceleration of falling objects did Galileo resort to his famous experiments in which he measured the rate of acceleration of objects rolled down an inclined plane. In other words, the tests he conducted under actual conditions served merely to confirm his antecedent "thought experiment" (Westfall 1977, 16–24).

Galileo boasted of his achievement of arriving at his mathematical law of acceleration, but others, long before him, Nicole Oresme (ca. 1320–82), in particular, had already discovered it, and, by the late sixteenth century, it was widely known. The beginnings of modern physics did not spring so suddenly from Galileo, as historians once thought, but had a long medieval history. Yet it was Galileo who first put Oresme's rule of motion to the test of experiment and determined that it could be applied to actual conditions (Cohen 1960, 107–13).

Galileo also showed that a projectile follows the path of a parabola because it has, simultaneously, a combination of two independent motions—a uniform motion in a horizontal direction and a uniformly accelerated motion downward. The independence of the two motions can be illustrated in a drawing (Fig. 9). In the same intervals, a pro-

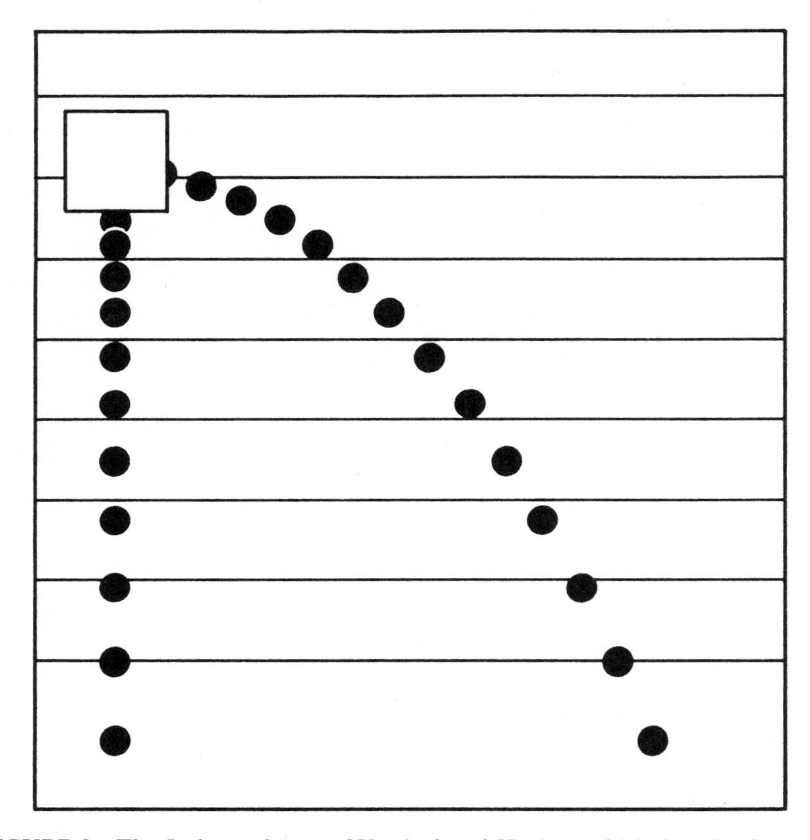

FIGURE 9 The Independence of Vertical and Horizontal Motion. In the same time intervals, the projected ball falling along a parabolic path drops the same distance as the ball falling vertically.

jected ball following a parabolic path drops exactly the same distance as the ball allowed to fall vertically. The uniform horizontal motion is, to some extent, an anticipation of inertia, a concept fundamental to modern physics and fully developed later by René Descartes (1596–1650) and Isaac Newton (1642–1727). Inertia is the tendency of a body that is moving in a straight line at a constant speed to continue to do so, even though it loses physical contact with the original source of that motion. It is an idea that breaks decisively with the Aristotelian insistence that a body cannot move without a constantly propelling force external to it.

But Galileo is a long way from the idea of linear inertia first fully enunciated by Descartes. Let us see why. For Galileo, following Copernicus, the cosmos is a feat of divine engineering built on the model of the circle. Galileo subscribed to the ancient belief in circularity as the fundamental principle of cosmic order. (One sees why he found Kepler's elliptical orbits unacceptable.) The only perpetual motion he could therefore countenance was the circular motion of the planets around the sun. So long as they neither approach the center nor recede from it, no cause operates to change their velocity or to alter their natural places (Westfall 1977, 24). Thus, the only kind of inertia Galileo could conceive of was circular. Linear inertia was ruled out of court. As he said,

> straight motion being by nature infinite (because a straight line is infinite and indeterminate), it is impossible that anything should have by nature the principle of moving in a straight line; or, in other words, toward a place where it is impossible to arrive, there being no finite end. For nature, as Aristotle well says . . . , never undertakes to do that which cannot be done, nor endeavors to move whither it is impossible to arrive. (quoted in Cohen 1960, 125)

In the *Dialogue*, Galileo rejected the Aristotelian distinction between celestial and terrestrial physics but "clung to a distinction between naturally occurring motions which are uniform and circular, and forced ones which are accelerated and rectilinear" (Hall 1981, 49). For all his new thinking about the physics of motion, Galileo did not break free of the ancient grip of circularity. In the *Dialogue*, he did not see that the planets circling the sun are not behaving purely inertially; rather, their linear inertia is being constrained by the force of gravity. This is what Newton would show some fifty years later.

Scientific Heresy

Galileo's works had tremendous impact not only on science but on religion. His defense of Copernicanism contradicted some passages in the Bible that implied that the sun moves around a stationary earth. In the wake of the Protestant Reformation, the hierarchy of the Catholic Church feared that Galileo's science would encourage people, educated and uneducated alike, to interpret the Bible for themselves and thus spread false belief. The Council of Trent (1545–63), a meeting of Catholic leaders designed to strengthen the church against the threat of Protestantism, had created the machinery for censoring books and punishing heresy, and many thinkers, like Bruno, had suffered from the ensuing repression at the hands of the Inquisition. But many Catholics also agreed with the biblical scholar Cardinal Cesare Baronio (1538–1607), who claimed that the Scriptures teach "how to go to heaven, not how the heavens go," and that therefore the Bible should not be used to decide scientific questions at all. In 1616, the Catholic Church, nevertheless, chose to take a harder line and forbade the teaching of Copernicanism. But church censors, persuaded by Galileo's supporters, some in high places, permitted Galileo to publish his *Dialogue* in early 1632. By August, however, the conservatives had seized the initiative; the book's publication was suspended and unsold copies confiscated. The following April, Galileo was put on trial by the Roman Inquisition, found guilty of heresy, and forced to declare that his Copernican teachings in the *Dialogue* were false and heretical and that the Ptolemaic system was correct after all. Placed under a life sentence of house arrest and forbidden to publish any further on Copernicanism, Galileo spent the next five years working on his new physics and writing perhaps his greatest book, the *Discourses on Two New Sciences,* which was smuggled out of Italy and published in Leiden (Holland) in 1638. He died in 1642.

The trial, which included a threat of torture, and condemnation of Galileo sent shock waves throughout the European republic of letters because, by 1633, Galileo not only was Europe's most famous scientist but his investigations provided a new and powerful model of what science could achieve and by what methods. And now he had been forced to kneel in an act of public abjuration before the authority of the church (Langford 1971; Redondi 1987; Shea 1986).

His books continued to be printed and translated outside of Italy and had a lasting impact on scientific development. What was so impressive about his thinking? It is often said that, harking back to Pythagoras, he geometrized space or nature; he showed, that is, that things are to be studied not in terms of their Aristotelian essences but in terms of their quantifiable features. Phenomena, moreover, can be known, with certainty, to the extent that they can be understood mathematically. As Galileo put it in *The Assayer* (1623):

> Philosophy is written in this grand book, the universe, which stands continually open to our gaze. But the book cannot be understood unless one first learns to comprehend the language and read the letters in which it is composed. It is written in the language of mathematics, and its characters are triangles, circles, and other

geometric figures without which it is humanly impossible to understand a single word of it; without these one wanders about in a dark labyrinth. (quoted in Drake 1957, 237)

Science, to be true, must abstract from nature all purely sensory perception of it. What is left—size, shape, weight, speed, direction—can be reduced to cold numbers. Here is the beginning of the distinction between the so-called primary and secondary qualities of physical objects, a distinction that would play an important role in the Scientific Revolution. Primary qualities (size, weight, and so on) are thought to be real because they are susceptible to mathematical measurement. Secondary qualities (sights, sounds, and so on) may derive from the primary ones but refer to our perceptions of things and not to things themselves, and our perceptions of such qualities are notoriously untrustworthy. Of these secondary qualities, Galileo wrote in 1623: "Hence I think that tastes, odors, colors, and so on are no more than mere names so far as the object in which we place them is concerned, and that they reside only in the consciousness" (quoted in Drake 1957, 274). Here Galileo is scoring a skeptical point similar to the one we saw Montaigne making in his *Essays* some fifty years before. It was a point that would not be lost on the new generation of French philosophers—Mersenne, Gassendi, and Descartes—to be considered in the next chapter.

Princely Science: The Pursuit of Knowledge and Power

Aristotle and Plato had put a high value on the contemplation of abstract truth, and that outlook, far from dying, had been revived by the medieval Scholastics. They had also insisted that truth emerges from properly conducted academic debate and disputation. This tradition persisted in early modern Europe. What also persisted was a deeply ingrained prejudice against the manual arts of artisans and peasants and a corresponding prejudice in favor of the liberal arts of academic education, in which gentlemen and those who aspired to be gentlemen were prepared for careers in the church, the law, and medicine.

In the sixteenth century, a new idea of science emerged to take its place alongside those already offered. This was the idea that science should be less about the contemplation and demonstration of old truths than about the discovery of new ones, even if this meant manual operations, going directly to nature, and getting one's hands dirty in the bargain. Paracelsus held this view, as we saw; in fact, he was one of its most influential exponents. For those who subscribed to this view, science came to be construed as "a hunt—as an attempt to penetrate territories never known or explored before" (Rossi 1970, 42). The thinkers involved redrew the boundaries between what educated people thought it is legitimate to know and what it is not.

Beginning with St. Augustine of Hippo (354–430), Christian thinkers regarded insatiable intellectual curiosity as a lust and a disease, a view revived and adopted by Montaigne. Augustine called curiosity an "empty longing" that leads us to pry into nature's secrets, "which it does no good to know" (quoted in Eamon 1991, 33). Others went further and held that God prefers us not to inquire into the processes of creation and that, if we persist in doing so, we may overstep the boundaries that God has set and pursue forbidden knowledge. But, in the sixteenth century, curiosity about the secrets of nature, from being a vice, was made a virtue in various contexts—by Paracelsus, for instance, and at the courts of princes. In this rarified, courtly setting, curiosity, the ardent pursuit of knowledge, and, especially, the unlocking of nature's secrets came to be celebrated for two reasons—one practical and the other ideological.

First, the new rulers needed technical expertise to be able to carry out their projects—the building and decoration of palaces and gardens, the planning of cities, and, especially, the preparation for war, which, driven by the early modern arms race, required the latest in fortifications and weaponry. So rulers placed high value on those who had the technical know-how that could be translated into practical results. There were many who, claiming to have the necessary skill, answered the princes' call. The rewards could be considerable. To take an extreme case, Giambattista Della Porta (1535–1615) said that his princely patrons paid him more than 100,000 ducats (gold coins, especially ones minted in Venice) for his services as "the foremost hunter and divulger of secrets" of his day (Eamon 1991, 40). Even Galileo was not immune. To win Medici patronage, he let it be known at court that he had, as he put it, "particular secrets, as useful as they are curious and admirable," which he was willing to reveal in return for a suitable reward (Eamon 1991, 37).

Second, the princely cult of curiosity was driven by a political and religious agenda. Rulers were obsessed with their reputations and did whatever they could to advance them. One route was to surround themselves with thinkers whose claims to secret knowledge or magical power could add luster to the court. If they practiced magic, it was the so-called natural magic that derived from the theories of Ficino and Paracelsus. Typically, such in-house philosophers collected medical and chemical recipes that might either be published or kept for the private delectation of the patron and his friends. Sometimes these recipes were tested to determine whether they worked or not, and procedures were set up for conducting simple experiments and establishing efficacy. Rulers, for their part, sponsored scientific meetings, or "academies," and built libraries, research facilities, and cabinets of "curiosities" as many expressions of their power. They patronized alchemical adepts, like Della Porta, astrologers and astronomers, like Kepler, and astronomers and mathematicians, like Galileo. Several Medici princes and the Holy Roman Emperor Rudolf II (1576–1612) studied these arts for themselves. In doing so, they cultivated the Renaissance image of the learned prince who combines knowledge and power and whose knowledge, by giving him control over nature, enhances his power. This courtly ideal of science, it seems, was directed less to serving the common good than to reflecting the prince's glory (Eamon 1991; Westman 1980, 117–27).

Around Rudolf's court at Prague swarmed a horde of magicians—some Paracelsians, some Brunonians, including Bruno himself for a short time—and all dedicated to unlocking the secrets of nature. Their goals echoed those of the late Renaissance magicians, like Bruno and Paracelsus: The discovery of nature would overcome the religious divisions and usher in a world reformation. The magicians not only looked to Rudolf for patronage but pinned their hopes on him to lead Europe toward universal prosperity, order, and harmony, if not the millennium itself (Evans 1973, 198–284). Many of the themes of this princely cult of investigation into nature—the courtly aspect, the link between knowledge and power, the drive for dominion over nature, the organization of research, the millenarianism—would be echoed in the writings of Francis Bacon. But, in his version, the magical elements were filtered out in favor of a more sober pursuit of nature's secrets. For Bacon, moreover, the discovery of those secrets was meant to serve both the prince's glory *and* the public good (Kaufmann 1993, 174–94).

Bacon

The sixteenth century was the golden age of the advice book. Despite low rates of literacy, such books flooded the market. There were manuals claiming to give practical information on myriad subjects, from elite concerns, such as how to rule a state (for example, Niccolo Machiavelli's *Prince*, 1532) or how to cut a figure at court (Baldassare Castiglione's *Book of the Courtier*, 1528), to more popular subjects, such as how to cast a horoscope or work a cure. The printing press and the book market functioned as an early-day information superhighway; everyone, it seemed, had a secret to tell and to sell. Bacon, at the end of the century, was the child of that tradition. The advice books he wrote made the most ambitious claims of all, for they sought to teach us nothing less than how to reform thinking itself so that we might go on to understand and control nature for the common good (Febvre and Martin 1984; Hall 1962, 17–67, 197–212, 247–64).

Francis Bacon (1561–1626) was an important English humanist and high civil servant who worked under two successive monarchs, Elizabeth I (1558–1603) and James I (1603–25). As Galileo's contemporary, Bacon formulated a very different, but no less influential, idea of science. He had read Paracelsus and the Paracelsians, Montaigne and the skeptics, and perhaps Bruno, and responded to their thinking both positively and negatively. For Bacon the pessimist, the human mind is "an enchanted glass, full of superstition and imposture, if it be not delivered and reduced." But for the optimistic Bacon, the mind is also "the lamp of God," capable of learning and knowing, and nothing in the world "is denied man's inquiry and invention" (Bacon 1955, 162, 295). Bacon agreed with the skeptics that the mind can be led astray into error and vice and has, in fact, been so misled by both Paracelsians and Aristotelians, but he also held that effective remedies are available.

New Aims and Methods

Bacon's attack on Aristotle and Paracelsus is both intellectual and ideological. They are both system-builders whose ideas are expressed in high-flown, abstruse jargon that fails to say anything reliable about the natural world it pretends to explain. Neither Aristotelians nor Paracelsians, moreover, are successful at achieving useful results because the former sell the human intellect short while the latter do the opposite and overreach its limits. The goal of the Aristotelians is contemplation without any thought of practical application. They are further hampered, in this respect, by their cherished view that, as Bacon says, "nothing difficult, nothing by which nature may be commanded or subdued, can be expected from art or human labour," because they think that "whatever has not yet been discovered and comprehended can never be discovered or comprehended hereafter" (1955, 507).

At least in this respect, Bacon says, Paracelsus is an improvement on Aristotle. According to Paracelsus, God has designed nature in such a way that gifted people, the magi, can find out all its secrets and use them to relieve suffering and poverty. Bacon may be said to have adopted Paracelsus's goal—the control of nature for the common benefit—while he rejected his methods. For both men, the truths of nature are not merely to be contemplated for their own sake, as Aristotelians held, but to be put to work to alleviate human want (Rossi 1968, 32).

If Bacon borrowed fundamentally from Paracelsus as far as goals are concerned, he sharply departed from him on the question of how to reach those goals. Paracelsus, we recall, had declared that God gives some people, the magi, the power to "perform miracles on earth . . . , for man is of divine nature." But, for Bacon, this was to claim far too much for anyone's powers, even the most gifted, and he called Paracelsus a "sacrilegious impostor" for making such extravagant claims (Farrington 1964, 66). Such conceit, Bacon said, had misled the Paracelsians in their study of nature. They collected too little data, made too few experiments, and then leaped to conclusions on this inadequate foundation. Their language was obscure, their findings incommunicable. They claimed to be divinely inspired and made a fetish of secrecy, which was a cover for error, imposture, and simple greed—their desire, that is, to sell their "cures" to the sick and gullible (Farrington 1964, 52–3).

For Bacon, science, properly conceived, would correct for the shortcomings of the Paracelsians and the weaknesses and self-deceptions even clever thinkers fall victim to, what Bacon called "the Idols of the human mind." As he said, "It is a false assertion that the sense of man is the measure of things. On the contrary, all perceptions . . . are according to the measure of the individual and not according to the measure of the universe" (Bacon 1955, 466, 470). Montaigne had already said something similar to this, and Galileo was soon to say it again, as we saw. The effect of their skepticism was different in each case. Montaigne had retreated into a private world of self-improvement; Galileo had discovered that knowledge was still possible, that truth lay in the cold certainties of mathematical physics—number, weight, measure, and motion.

Bacon, like Galileo, was an optimist; scientific knowledge is attainable. But he charted a different course—not mathematics but induction—collecting information, conducting experiments, and interpreting their results. Galileo, too, had used experiments, but, as we saw, his experiments served to confirm general conclusions already arrived at mathematically. Bacon reversed the process: General conclusions would be reached by collecting data and using them to design experiments. To Bacon's mind, this would be a slow, patient process involving collaboration among many investigators, who would set about observing nature and compiling records of those observations, from which useful knowledge might eventually be derived: "Nature," he said, "cannot be conquered but by obeying her." In the process, the idols of the mind must be defended against and warded off to every possible extent. "Men's wits require not the addition of feathers and wings, but of leaden weights." The mastery of nature will come from "a true and proper humbling of the human spirit" (quoted in Farrington 1964, 93, 119, 133).

Science must also be a collaborative, public endeavor, directed, administered, and paid for by the monarchical, bureaucratic state (about which more below). The resulting knowledge, though subject to state control, should not be kept secret or sold to the highest bidder, as Bacon accused the Paracelsians of doing, but should be made available for the common good, as defined by the appropriate public authority. Such knowledge would also be that much more genuine and useful for being the result of scientific collaboration. This was in sharp contrast to the secretive practices of the Paracelsians, who claimed to be divinely inspired and expected the world to take their word for it. How could their knowledge claims be trusted? But the Baconian product came with the warranty of collaboration.

Optimism

One influential feature of Bacon's idea of science was his guarded but robust optimism. While he tried to protect against what, to him, were the rash, spurious, and counterproductive claims of the Paracelsians, Bacon made some exuberant claims of his own, which would prove remarkably contagious in both England and France. The source of this Baconian optimism was his take on original sin. Echoing Paracelsus and Bruno, Bacon subscribed to a version of the idea of the Fortunate Fall:

> Man by the Fall fell both from his state of innocence and from his dominion over the creation. Both of these losses can, however, even in this life, be in some measure recovered, the former by religion and faith, the latter by the arts and sciences. For the creation was not by the curse made altogether and forever rebellious to man. (quoted in Farrington 1964, 29)

That biblical curse, the curse of labor—"In the sweat of thy brow thou shalt eat bread"—turns out to present a golden opportunity (Low 1985, 131–42). According to Bacon, if humankind commits itself to one particular kind of labor, namely, the study of nature by proper Baconian methods, it will "recover that right over nature which belongs to it by divine bequest" (1955, 539). By good Baconian science, "but assuredly not by [Aristotelian] disputations nor by [Paracelsian] vain magical ceremonies, the creation is at length and in some measure being subdued to supplying of man with bread, that is with the satisfaction of his human needs" (Farrington 1964, 29). At times, Bacon's optimism can sound utopian. Those who follow his path to the discovery of nature will become "a blessed race of Heroes or Supermen who will overcome the immeasurable helplessness and poverty of the human race, which cause it more destruction than all giants, monsters, or tyrants, and will make you peaceful, happy, prosperous, and secure" (Farrington 1964, 72). There is also evidence to suggest that Bacon was a millenarian, that he thought that his science might help to usher in the millennium of biblical prophecy that so many Protestants, at the time, were expecting (Bacon 1955, 512–3; Farrington 1964, 131–2). Certainly his followers in seventeenth-century England put a millenarian spin on his teachings (Trevor-Roper 1967; Tuveson 1949; Webster 1975, 1–31).

Bacon was insistent on one point: The pursuit of knowledge must be undertaken in the right spirit and for the right reasons, "not . . . for superiority to others, or for profit, or fame, or power, or any of these inferior things; but for the benefit and use of life" in a spirit of charity, of which "there can be no excess" (1955, 437). Paracelsus had said something similar, but, for him, charity is a virtue of simple folk, peasants, and artisans, while, for Bacon, it has no specifically lower-class associations at all and has been purged of any Paracelsian ones. In fact, vanity and greed, the Paracelsian vices in Bacon's view, are antithetical to it. Charity, as he made clear, is not a selfless love that denies legitimate self-interest and sacrifices the individual on the altar of the common good. Rather, it is a virtue founded on the assumption that "there is formed in everything a double nature of good"—individual and collective. The collective is superior to the individual, but one should not pursue either at the reckless expense of the other: "Divide with reason," Bacon advised, "between self-love and society" (Tuck 1993, 113; Bacon 1955, 34–5, 63).

It is Christian charity, nonetheless, that will point science in the right direction, that is, toward serving not the selfish interests of one or a few but the common good. The benefits of knowledge will thus be maximized, and there is also an intellectual reward. If science is undertaken in a generous spirit rather than being driven by the vanity and pride of the practitioners, they are less likely to be carried away by their own fancies and more likely to achieve "the true and legitimate humiliation of the human spirit" so necessary to the discovery of truth. The honest seeker of knowledge is, in a word, less likely to submit to selfish emotions, whether "weak fears or vast desires," and more likely to attend to careful observation of the natural world. Charity is a moral virtue, which, it turns out, has beneficial consequences for both the methods and goals of science (Bacon 1955, 432–7, 162–5).

For Bacon, there is never any question of a conflict between his science and right religion, not because they exist in separate compartments but, on the contrary, because science itself, if done properly, is a profoundly Christian adventure. Not only does it rescue human beings from some of the ravages of the Fall and add, as nothing else can, to their "power and greatness," but genuine science contributes directly to our moral and spiritual development. It displays the power of God in the creation and, as such, is "given to religion as her most faithful handmaid." After the Holy Scripture, science is "at once the surest medicine against superstition, and the most approved nourishment for faith." Of course, the only science that can provide such support to religion is Baconian empiricism, that is, collecting information, performing experiments, and, all the while, guarding against the idols of the mind. By contrast, the Paracelsian "unwholesome mixture" of science and religion produces "not only a fantastic philosophy but also a heretical religion" (Bacon 1955, 527, 509, 484).

The study of nature also provides moral instruction. Bacon insisted that such an education combine ethical theory with practice and, through practice, the engendering of good habits. For him, acquiring natural knowledge and putting it to use for human benefit have exactly this effect. Scientific husbandry undertaken in a spirit of charity, for example, is an ethical activity in and of itself and also teaches morals; it makes us good as we do good (Bacon 1955, 318–9). Elsewhere Bacon said, more generally, that learning pays off in moral improvement: "Nay further, in general and in sum, certain it is that *veritas* [truth] and *bonitas* [goodness] differ but as the seal and the print; for truth prints goodness" (Bacon 1955, 215–6; Rossi 1968, 109–10).

But the road is not smooth. According to Bacon, human beings, left to their own devices, "are full of savage and unreclaimed desires, of profit, of lust, of revenge." The common people are the worst; echoing Luther, he said that their "innate depravity and malignant disposition," if left unchecked, would dissolve everything "into anarchy and confusion" (1955, 202, 409). But there is a remedy:

> And it is without all controversy that learning doth make the minds of men gentle, generous, maniable and pliant to government; whereas ignorance makes them churlish, thwart and mutinous: and the evidence of time doth clear this assertion, considering that the most barbarous, rude and unlearnèd times have been most subject to tumults, seditions and changes. (Bacon 1955, 170–1)

Here Bacon updated an argument drawn from ancient Roman thinkers, especially Cicero, and revived by many Renaissance humanists. They held that civilization, in both its origins and its perpetuation, is crucially dependent on great leaders. These are the individuals who have the reason and the eloquence to persuade whole populations, mired in conditions of barbarism, to give up their wild and savage ways and to set to work to build a peaceful and prosperous future (Hale 1994, 355–72; Tuck 1979, 33–4). Bacon cleverly altered this ancient argument by claiming that scientists deserve even more credit than statesmen for pacifying and civilizing the human race, in part because scientific improvement is peaceful, whereas "the reformation of a state in civil matters is seldom brought in without violence and confusion" (Bacon 1955, 538–9). It is, he said, his own kind of collaborative science that "teaches the peoples to assemble and unite and take upon them the yoke of laws and submit to authority, and forget their ungoverned appetites . . . ; whereupon soon follows the building of houses, the founding of cities, the planting of fields and gardens with trees" (Bacon 1955, 412).

Science, for Bacon, helps to foster moral and social order. This is a message, as we shall see, that would not be lost on his followers in seventeenth-century France and England. Bacon, however, was no Pollyanna. To be sure, he hoped for and worked for material and moral improvement, "the advancement of learning," as he called it; but he did not believe in inevitable progress forever and in a straight line. Echoing an ancient Greek tradition revived by the humanists, he held that history is an endless cycle of good times and bad. The historical process is a dynamic one; growth and prosperity sow the seeds of their own destruction, and out of this decay comes rebirth (Bacon 1955, 412–3; Tuveson 1949, 56–67).

Royalism and Utopianism

If Bacon's legacy is deeply Christian, it is also deeply statist. He was educated in the English humanist or "commonweal" tradition that gave the monarchical state an active role in increasing the wealth of the kingdom, reducing widespread poverty, and solving the attendant social problems of vice and crime. He was also following in the footsteps of his uncle and father, both important officials in the government of Elizabeth I, officials who strove to implement these reforming policies. Bacon's thought was crucially shaped by his response to developments he regarded as politically dangerous and a serious challenge to the stability of the realm. "In Bacon's eyes, natural philosophy could be . . . refashioned into a splendid support for the Tudor state." What had become politically dangerous, in his view, was a subculture of Puritan (Calvinist) gentry, yeomen, and artisans fueled by three ideological ingredients—a religious Puritanism increasingly hostile to an apparently unreformable state church, a Paracelsian medical literature insisting that healing knowledge does not come from the academically trained physicians and clergy but from the unmediated experience of nature, and a flood of printed tracts aimed at the Puritans and furnishing advice on how to improve their lands and incomes. Here were the makings of "voluntary communities" working independently of church and crown, a sinister state within a state, representing a "commonweal of the godly" rather than "the official Tudor ideal of an all-embracing, national community." Bacon found all this so subversive that he went on to construct his natural philosophy partly in answer to it. Instead of the private knowledge and separatism of the Puritan elect, he proposed a program of state-managed knowledge that would be profitable not to the few but to the many, not to Puritans alone but to the whole kingdom (Martin 1992, 45–71).

Something of what Bacon envisaged can be found in his *New Atlantis* (1626), an account of Bensalem, a utopian or ideal kingdom in which science plays a key role. The citizens of Bensalem are prosperous and happy, if also docile and passive. It is a paternalistic and hierarchical state in which decisions are handed down from above. The behavior of the common people watching an official procession in the street is compared to that of an army "in battle-array." The key to the success of Bensalem is Salomon's House, "the lantern of this kingdom," founded some nineteen hundred years before, named after the biblical Solomon, "the King of the Hebrews," and dedicated to the study of nature. Its elite members are "to discern between divine miracles, works of nature, . . . impostures and illusions of all sorts." Thus, one of its jobs is to maintain the proper relation between science and religion and to guard against that "unwholesome mixture" of the two, which Paracelsianism produced. King Solamona, who founded Salomon's House, "had a large heart, and was wholly bent to make his people happy." His latter-day successor is an equally good Baconian: "He had an aspect as if he pitied men." Bacon's utopia is shot through with his brand of Christian charity—paternalistic, if not patronizing, and carefully marshaled by the state (Bacon 1955, 573, 554, 561, 572).

The science preached is also Baconian in both its experimental methods and its practical goals. Like everything else about Bensalem, scientific activity is structured according to a rigid hierarchy. There is a carefully graded division of labor running from the lowly observers and slightly more elevated compilers to those who oversee the process. Bacon's science is meant to serve the common good and to answer those who, like the Paracelsians and Puritans of the day, would use knowledge for their private ends. But, like theirs, the state science of Salomon's House is explicitly elitist and secretive: "We have consultations, which of the inventions which we have discovered shall be published, and which not: and take an oath of secrecy, for the concealing of those which we think fit to keep secret." Big Brother is not the invention of the twentieth century. More interesting, Salomon's House may not even be answerable to its putative master: Regarding the research findings, which are kept from the people, "some of those we do reveal sometimes to the state, and some not" (Bacon 1955, 582–3).

Bacon's scientific utopia was only one of many written throughout Europe during the early seventeenth century, but his was one of the most influential. It described an ideal world of order, peace, and prosperity that had wide appeal in the midst of the Thirty Years' War (1618–48) and in the aftermath of the European wars of religion. Bacon's vision was of science soberly pursued by an intellectual and social elite, a science shorn of Aristotelian disputation, productive of much more heat than light, and the heretical magic of Paracelsus and Bruno. Bacon's vision of science subject to state control appealed to rulers; his vision of science applied to alleviating hunger and disease appealed to everyone, especially advocates for the poor and dispossessed. The conquest of nature for the common good and the alliance of knowledge and power, these were potent messages, and they traveled far (Brown 1934; Solomon 1972, 74; Tuck 1993, 285).

The Revolution in Medicine (1543–1661): Vesalius to Malpighi

Simultaneous with the revolution in physical astronomy which culminated in Newton's work, there was a revolution in medicine, with consequences, if less spectacular, perhaps no less profound. The high point in this revolution came with the discovery of the circulation of the blood by the English physician William Harvey (1578–1657). He received his medical education at the University of Padua, a medical faculty that was renowned as the best in Europe at the turn of the seventeenth century, when he went there to study. During the previous sixty years, its anatomy professors, especially Andreas Vesalius (1514–64), Matteo Realdo Colombo (ca. 1510–59), Andrea Cesalpino (1524–1603), and Harvey's teacher, Fabricius ab Aquapendente (ca. 1533–1619), were famous for their careful anatomical studies, and Padua's philosophers were perhaps equally distinguished for their subtle examination of medical theories and doctrines deriving from several ancient traditions. Harvey was the beneficiary of this expertise in both theory and practice.

Vesalius published the results of his anatomical research in his landmark book *De Humani Corporis Fabrica* (*On the Structure of the Human Body*) in 1543. The book, notable for both its detailed descriptions and its lavish and meticulous illustrations, set a new standard for clarity and accuracy in anatomy and made all earlier work outdated (Fig. 10). Colombo discovered the movement of the blood from the heart to the lungs and back again to the heart. Cesalpino expounded this blood flow and coined the term "circulation." He also spoke of the continuous return of venous blood and its outflow through the arteries but never demonstrated it as Harvey was to do. Finally, Fabricius demonstrated to his students, including Harvey, the valves of the veins but did not understand their true function.

The standard sixteenth-century teaching on the subject of the heart and the blood derived from Galen. There was no place in the Galenic view of the blood for its circulation. Rather, it was created in the liver, thought by Galen to be the principal organ of the body, from a juice of predigested food, called chyle, manufactured in the stomach and carried to the liver. Blood, once formed in the liver, then flowed out through a network of veins that reached every part of the body and provided it with life-giving nutrition. The blood, thus transported by the veins, was literally used up in the very process of providing a steady stream of nutrition.

FIGURE 10 Andreas Vesalius, medical doctor and anatomy professor, stares out at us from the center of this illustration. As he lectures, he performs a dissection (rather than leaving it to a surgeon-assistant), thus uniting theory and practice and overcoming the ancient separation of head and hand (and the ancient prejudice against handwork). (From Vesalius's *On the Structure of the Human Body*, 1543. Courtesy of Van Pelt Library, University of Pennsylvania.)

In this process, as Galen understood it, the heart also played a part. Some of the venous blood flowing from the liver reached the right ventricle of the heart, where either of two things happened to it. Most of it was relayed from the heart to the lungs by the pulmonary vein (our pulmonary artery), where it was also used up because the lungs, like the rest of the body, needed to be fed. The small amount of blood that did not go to the lungs was said to pass from the right ventricle into the left ventricle through invisible pores in the tissue (the interventricular septum) separating the two chambers. There, it was mixed with air received from the lungs by the pulmonary artery (our pulmonary vein), and this new mixture was then delivered by the aorta to the network of arteries that reached every part of the body. The purpose of this new mixture was to supply vital heat to the whole body, a process thought to be just as essential to life as nutrition, and this arterial blood was also said to be completely consumed once it reached its destination, just as the venous blood was, and not recirculated. In fact, there was no place in this system for any idea of circulation at all (Lindberg 1992, 125–31).

Colombo, who had been Vesalius's student at Padua, published *On Anatomy* in 1559 and announced a crucial discovery, the so-called lesser, or pulmonary, circulation of the blood, according to which, blood flows out of the right ventricle of the heart to the lungs and back to the left ventricle. So what is carried from the lungs to the heart, in other words, is not air, as Galen had held, but blood. Over many years, before and after 1616, Harvey built on this discovery in several stages, which we shall reduce here to a simple two-step process. First, he developed techniques for studying living, beating hearts, which meant using dying animals whose heartbeats were slow enough to permit close examination. The result was that his knowledge of how the heart actually works advanced enormously, and he could see it as a pump, taking blood from the veins and, with each muscular contraction, expelling it into the aorta. Second, and more decisively, he calculated the amount of blood actually coursing through the heart from the veins to the arteries. He gradually realized that the rate of transmission is such that the total blood supply would soon be depleted, the veins emptied out, and the arteries choked in no time at all, if the blood flowed in one direction only.

It was on this numerical basis that Harvey abandoned the Galenic theory of a one-way venous transmission and adopted the idea of the uninterrupted circulation of *all* the blood from the veins to the arteries through the heart. He published his findings in *De Motu Cordis* (*On the Motion of the Heart*) in 1628. The crucial proof for his theory, however, had to wait until 1661, when Marcello Malpighi (1628–94) announced his discovery of capillary blood vessels, linking arteries and veins, in the lung tissue of frogs. These capillaries, invisible to the naked eye, could be clearly seen under Malpighi's improved microscope, an instrument that would have as much impact as the telescope. (Galileo had devised a primitive microscope by 1610 [Bylebyl 1972; Wear 1990; Hall 1981, 166–74; Van Helden 1983, 71].)

Harvey was an Aristotelian, though not a slavish one, and, in adopting his theory, he was probably influenced by the crucial role circularity played in Aristotelian cosmology. Indeed, Harvey seems to have assimilated the circulation of the blood to the cosmic kind: Just as circular motion is, according to the Aristotelians, the principle of order and harmony in the heavens, so is it the principle of perpetual renewal in the bodies of animals. The blood, according to Harvey, is essential to life in both a material and spiritual sense. "It is," he said, "celestial, for nature, the soul, that which answers to the essence of the stars, is the inmate of the spirit, in other words, it is something analogous to heaven, the instrument of heaven, vicarious of heaven" (quoted in Westfall 1977, 91). For Harvey, following Aristotle, nature, including the heart and the blood, operates teleologically through its forms and correspondences and not from mechanical principles or by mathematically measurable forces. But Harvey was also shrewd enough, as we have seen, to put mathematics to effective use, when (as in measuring blood flow) it served his purposes.

It is ironic that the medical revolution Harvey made should owe so much to a revitalized Aristotelianism just at the time that the ancient Aristotelian world-picture was being torn down and replaced by a small cadre of determined anti-Aristotelians. In fact, one such extremist, René Descartes, even went so far as to fasten on Harvey's Aristotelian physiology only a few years after it was published, translate it into the terms of his own mechanical philosophy, and thus appropriate it for service in the anti-Aristotelian crusade (Descartes 1968, 65–72). But even so, what cannot be denied is that, by making a new start in the study of living creatures, Harvey the Aristotelian turns out to be just as revolutionary as Galileo, Bacon, or Descartes (Westfall 1977, 86–94).

Science in Seventeenth-Century France

PHILOSOPHICAL DEVELOPMENTS IN France from 1610 to midcentury proved crucial to the origins of modern science. All the intellectual currents we have so far traced out, and some new ones, made themselves felt in this French, and, especially, Parisian, context, and the resulting debates pushed science in important new directions. This chapter will first describe the main parties to these debates and then move on to examine the principal outcomes.

The Heretical Challenge

Montaigne, we recall, had used his skepticism about our capacity for achieving certainty as a foundation for his fideism. If we cannot arrive at the truth by reason and the senses, then we should accept the religious truths of the Catholic Church on faith, that is, as so many revelations from God. Others in the seventeenth century continued to follow him in this regard. But, as we have also seen, Charron turned skepticism against revealed religion and argued that, while it might be used by government to control the gullible masses, it is unworthy of the perspicacious few who, seeing it for the pious fiction that it is, must free themselves from its grip and live up to their status as rational individuals. Skepticism, in the right hands, might breed a virulent secularism.

Equally dangerous to orthodox religion were the naturalists. Since the Middle Ages, radical Aristotelians had refused to accept the synthesis, worked out by St. Thomas Aquinas, of revealed religion and Greek philosophy. One of the most influential of these subversives was

Pietro Pomponazzi (1462–1525), who could not reconcile Aristotle, for whom the human soul was mortal and the world eternal, with the Christian doctrines of a created world and immortal human souls. Pomponazzi concluded that these church teachings, failing the test of reason, must be accepted on faith. This view was consistent with Pomponazzi's naturalistic approach to explaining the world in general. For him, Aristotle provided the philosophical means for understanding nature as a self-activating system not requiring supernatural agency to preside over it and keep it going. The existence of angels and demons and the validity of miracles are thus called into question. From a naturalistic perspective, the world may be deemed independent of God, or, alternatively, nature itself may be seen as divine, as Bruno, for instance, had done (Copenhaver and Schmitt 1992, 103–12; Dear 1988, 24; Schmitt and Skinner 1988, 653–60).

Pomponazzi's naturalism, condemned by the church in 1513, spread from Italy in the sixteenth century and was taken up by the Carmelite monk Giulio Cesare Vanini (1585–1619), who studied in Italy and spent his later life wandering through western Europe seeking patrons and spreading heresy. He published two books, one in Lyons in 1615 and one in Paris the next year, in which he spelled out his views. He made the mistake of settling in Toulouse, "a town famous for its zeal in the suppression of heresy," where, lacking a powerful patron, his enemies closed in on him. He was tried and convicted for atheism and burned at the stake in 1619. A contemporary account claimed, "He professed to teach medicine, but in reality poisoned the minds of imprudent youths; he mocked at sacred things, vilified the Incarnation, knew no God, attributed all things to fate, adored Nature as the bounteous mother and source of all being" (Spink 1960, 29). Like Pomponazzi, Vanini denied that the immortality of the soul could be successfully defended by rational arguments. Like Bruno, he conflated God and nature while arguing (also like Bruno) that God cannot be reduced to mere nature (Hine 1984; 1976).

Four years after Vanini's execution, the poet Théophile de Viau was put on trial in Paris for views that echoed Bruno's and Charron's. Viau was representative of a larger group of literary men who scraped together a livelihood in the capital and often wrote in defiance of established norms and traditional values. Cynical and outrageous, they published licentious verse intended to shock the reading public. There was a market for such wares, and they were often printed without attribution. Viau narrowly escaped Vanini's fate; in 1625, he was convicted of atheism and condemned to be banished (Spink 1960, 42–4).

Besides the threat presented by these naturalists, Paris, in 1623, was the scene of a Rosicrucian scare. The Rosicrucians were reputedly an international secret society, the so-called Brothers of the Rosie Cross, devoted to discovering nature, healing the sick, and accomplishing the millennium. The actual existence of such an organization has never been confirmed by the evidence. What we have instead is a considerable literature, published mainly during the second and third decades of the seventeenth century, announcing its existence and discussing its goals. It was seen, in this literature, to be committed to realizing its aims through a revival of the Hermetic magic and alchemy of the previous century (Yates 1972).

The Rosicrucian "movement," wherever it appeared, had its zealous supporters and detractors and caused a sensation. In 1623, placards appeared on the streets of Paris announcing the arrival of Rosicrucian "deputies" who promised "to draw men from error and death" (Yates 1972, 103). By 1624, Rosicrucianism had enough public support to cause the government of Paris to forbid the teaching of new alchemical doctrines (Debus 1975, 29). The authorities were concerned because Rosicrucianism was loosely associated with heretics—Protestants, fanatical antipapists, and devil worshippers. It was being likened by its enemies to a kind of witchcraft. The putative brethren, all the more menacing for being invisible, became the object of popular paranoia, which led to a witch-hunt (Yates 1972, 103–17).

In 1616, the English Paracelsian physician Robert Fludd (1574–1637) published an apology for the Rosicrucians, aimed against their leading detractor, Andreas Libavius (1540–1616), who hated their extravagant claims and bombastic language, "the offspring," he said, "of Paracelsian fantasy" (Trevor-Roper 1985, 184). Fludd, for his part, mounted an all-out attack on the traditional educational curriculum and called for replacing it with alchemy, natural magic, Pythagorean number mysticism, and Paracelsian medicine (Debus 1975, 19–35). Fludd's writings prompted responses from leading philosophers and mathematicians. Kepler accused him of being "enigmatic and Hermetic" (Debus 1975, 26). Both Kepler and Fludd had been deeply influenced by the Pythagorean and Neoplatonic search for the keys to revealing nature's underlying mathematical harmonies. But Fludd, as his critics saw, put the cart before the horse; that is, his numbers and evidence were made to fit a preconceived picture of the universe. Kepler, in contrast, altered his hypotheses to accommodate his observations.

At the heart of Fludd's position was his belief that the Rosicrucians could arrive at a perfect knowledge of nature. As he said, "the alchemists and Hermetic philosophers . . . understand the true core of the natural bodies" (quoted in Debus 1975, 29). Like Paracelsus, he believed that certain individuals can be inspired to penetrate the secrets of nature. Like the Renaissance Hermetists, Ficino, Pico, and Bruno, he believed that such a revelation had occurred in remote antiquity and that this lost wisdom could be recovered for use in the present. As we have seen, Fludd was not alone in holding these views. There were many who believed that the time was ripe for a new alchemical and Hermetic revelation that would usher in the millennium and supply the religious and philosophical foundations for the ensuing order. Fludd and his ilk provided the blueprint for a new reformation, largely conducted outside the bounds of the old churches by those who saw themselves as constituting a new clergy inspired with power to control nature. Fludd himself claimed that "mystic and occult Chemistry" was "nothing else" but a kind of "practical Theology" (Debus 1975, 31–3).

In the wake of the Rosicrucian scare of 1623, Fludd's French critics spelled out the heretical implications of his position. For the friar and mathematician Marin Mersenne (1588–1648), Fludd was a "cacomagus, a teacher and purveyor of stinking and horrible magic." Mersenne was not opposed to alchemy but called for the establishment of alchemical academies to supervise research and punish charlatans (Debus 1975, 30–1). For Mersenne's colleague, the French priest and philosopher Gassendi (1592–1655), the logical outcome of Fludd's position would be to make "alchemy the sole Religion, the Alchemist the sole Religious person, and the tyrocinium [apprenticeship] of Alchemy the sole Catechism of the Faith" (Debus 1975, 33). Fludd proposed nothing less than to replace revealed Christianity with a new alchemical revelation, and his enemies knew it. Bruno had been executed for espousing his version of Hermetic philosophy. The response of devout Catholics like Mersenne and Gassendi to Fludd's Hermetism turns out to be far more complex and important.

The Natural Philosophers' Response

Mersenne and Gassendi reacted not only to Fludd and the alchemists and magicians but to the general intellectual crisis that had overtaken literate and scientific culture in early seventeenth-century France. The Rosicrucians represented one threat, but there were also the naturalists, pantheists, and radical skeptics, and they too had to be answered if Catholicism was to be upheld and science brought into proper relation to it. Because of Bruno's heretical teachings, for instance, Mersenne called him "one of the wickedest men who ever lived" (quoted in Yates 1964, 444–5). Nor was it good enough to fall back on an outmoded Scholasticism for solutions. Mersenne, Gassendi, and their circle had learned the lessons that Montaigne, Bacon, and Galileo had to teach. The first had shown that neither reason nor the senses can be trusted to deliver reliable knowledge. The second had mapped out a strategy for mitigating the deficiencies of reason and perception and for harnessing science to a larger project of conservative social reformation. And the last had shown the power of numbers, that is, the way to achieve mathematical certainty even in the face of observational uncertainty.

How, then, did Mersenne and Gassendi deploy these resources to meet the various challenges to what they regarded as orthodox religion and proper science? Their task was both philosophical and polemical. They had to discover a credible way to investigate nature, defeat heresy, and preserve, or even enhance, social order. All three aims and motivations were involved at the same time. Which ones were the more important—the social, the religious, or the scientific—is impossible to determine. What can be said is that they cannot be separated; they acted together.

Aristotelians had always said that the knowledge we get from the senses can be trusted if the sense organs are not diseased and are functioning properly. So, if we see that something looks red, then it is red in fact. As we have seen, Paracelsians and Rosicrucians claimed that truth can be revealed to some individuals by divine inspiration, and Hermetists held that the revealed knowledge passed down from Hermes Trismegistus represents privileged wisdom and is, especially, to be trusted, even revered. But Mersenne and Gassendi agreed with the skeptics that we can never know the real truth of things, whether by way of the senses or through divine or Hermetic channels.

Does this mean that it becomes impossible to do science, to obtain scientific information? If dogmatic science and absolute truth are the goals, the answer must be yes. But Mersenne and Gassendi argued for lowering the stakes and searching not for knowledge of things themselves but for knowledge of appearances. We may not be able to know what is really out there beyond our perceptions, but we can record, examine, test, and analyze those appearances and derive from this process some useful information. If we are willing to give up our claims to knowledge of truth and settle for knowledge of appearances, we can rescue science from the skeptics who say that nothing can be known at the same time that we silence the dogmatists who say that even the physical and metaphysical truth behind mere appearances can be known. As Mersenne said: "It is enough, in order to have certain knowledge of something, to know its effects, its operations, and its use . . . : we do not want to attribute to ourselves a greater . . . science than that" (quoted in Dear 1988, 40). The skeptics short-circuited the human capacity for natural knowledge, whereas the dogmatists made exaggerated and unrealistic claims for their perceptual and intellectual power. Avoiding both extremes, Mersenne and Gassendi sought to chart a middle way. This was the way of constructive or mitigated skepticism (Popkin 1979, 129–50; Tuck 1993, 285–94).

Mersenne

How, then, should the search for scientific knowledge be conducted? Mersenne and Gassendi both espoused a mitigated skepticism, but, beyond that, there were some important differences between them. Mersenne stressed the importance of mathematics in scientific investigation. Borrowing from St. Augustine of Hippo, Mersenne believed that God had created an orderly world based on mathematical ratios and proportions and expressing mathematical harmony. God alone can know this world directly, but, to the extent that the human intellect is illumined by God, we can use mathematics, God's language, to increase our knowledge not of things themselves but of the appearances of things. From patient and repeated observation and measurement of natural phenomena, patterns and regularities should emerge, which allow us to determine the probable (but not provable) causes of those appearances. Thus, as the historian Richard Popkin says, without being able to know whether the world "actually has the properties we experience . . . , we can develop sciences of appearances which have pragmatic value, and whose . . . findings are not doubtful except in a fundamental epistemological sense" (1979, 139).

From here, Mersenne went on to take only one more step, which was, in fact, a giant leap. He postulated that the world can usefully be conceived of as a machine whose workings can be explained mathematically. "Mersenne's mechanism, his world machine, was not set forth as the true picture of the real world, as it was for his fanatic friend Descartes, but as a hypothesis for organizing and utilizing our knowledge" (Popkin 1979, 140). Mersenne thus made two major contributions to the development of modern science: (1) his probabilism, the idea that from the study of appearances we can get probably reliable scientific knowledge, and (2) his hypotheticalism, the idea that conceiving the world as a machine can serve as a useful hypothesis whose testing, through experiments, may lead to probable certainties (Dear 1988). These two ideas would be developed further by his Parisian colleague Gassendi and, as we shall see in the next chapter, by the major scientific thinkers in seventeenth-century England.

Just as Mersenne's mitigated skepticism allowed him to trace the middle course between dogmatists and radical skeptics, so his version of mechanical philosophy, conceiving the universe as a vast cosmic machine, at least for the sake of argument, allowed him to answer Aristotelian naturalists, Hermetists, and magicians. For Mersenne, the world can be conceived as a machine created by a supernatural God to follow a course set by His providential care. He might intervene directly in the world and perform miracles, or He might resort to using lesser spiritual agents, angels and demons, to carry out His commands in the created order. But these agencies, whether God Himself or the angels and demons, would be supernatural powers and not the natural forces (material souls) postulated by Pomponazzi and Vanini. According to these radical Aristotelians, such forces pervade the universe and explain everything, both regular processes and what appear to be miraculous interventions, but are in fact merely inexplicable natural occurrences. Mersenne's mechanical hypothesis did not banish spiritual power from the world, but it did insist that such power is always supernatural in origin and so transcends and rules over the natural order rather than, as the subversive naturalists claimed, being immanent in it (Gaukroger 1995, 147–52; Lenoble 1943).

Mersenne's world-machine would also serve to answer the threat to orthodoxy posed by the Rosicrucians and Hermetists. There would simply be no room in a mechanical universe, set in motion by God and obedient to his sustaining Providence, for the paraphernalia of Neoplatonic and Hermetic magic, that is, the analogy between macrocosm and microcosm, stellar influences and ubiquitous angels and demons. Just as constructive skepticism might be used to address the heretical skeptics like Charron, so the hypothesis of a world-machine was used as a polemical device to oppose paganizing naturalism and various kinds of subversive magic. Pomponazzi and Vanini naturalized spirit; Bruno,

Paracelsus, and the Rosicrucians divinized nature. Against both heresies, Mersenne's mechanical hypothesis preserved an orthodox reading of the relationship between the spiritual and material orders, between God and nature (Lenoble 1943; Yates 1964, 432–40, 444–7).

Mersenne's science was polemical not only in its negative purpose—to defeat heresy—but also in its positive goals. It was designed to reduce rancor and division among philosophers and to produce a healing calm, a philosophical peace in which a community of thinkers might emerge devoted to a Baconian advancement of learning and civilization under the authority of the French king. Mersenne himself was known as Europe's postmaster general, so wide was his philosophical correspondence and so deep his commitment to building an irenic community dedicated to scientific discovery and communication and to Catholic enlightenment. Mersenne's was a vigorous attempt at an intellectual solution to a century of religious turmoil. From the ensuing philosophical peace, Mersenne believed, could come the political unification of Europe, under the leadership of the French monarchy, of course (Solomon 1972, 91).

Gassendi

In the story of European science, Gassendi occupies an even more important place than Mersenne. While sharing his mitigated skepticism, Gassendi developed the mechanical philosophy much further. From the ancients Lucretius and Epicurus and the modern Paracelsians, Gassendi borrowed an atomistic conception of matter. The fundamental building block of the material universe is the atom. Itself indivisible, the atom can combine and recombine in countless ways to constitute all the phenomena of nature. The physical universe is made up of atomic particles in motion. For Gassendi, their motion is not inherent in natural processes, as Bruno had held, but is initiated and guided by an omniscient Providence toward the achievement of God's purposes in the creation (Joy 1987; Westfall 1977, 39–42).

There was another way in which Gassendi made his atomistic and mechanical universe safe for orthodox Christianity. According to the ancient Greek atomists, the atoms are infinite in number and, as such, combine to form innumerable worlds. But, to conform to Christian requirements, Gassendi placed his atoms in a cosmic framework derived from Francesco Patrizi da Cherso (1529–97), who, in turn, had adopted it from the ancient Stoics. According to the Stoics and their medieval followers, we recall, our finite world is surrounded by an infinite void space. For Gassendi, like Patrizi, this finite world is not a plenum, as the ancient Stoics had held, but made up of particles moving through empty space. For Gassendi, moreover, these atomic particles are finite in number and created by God. But Gassendi's infinite void is another story, for it is both uncreated and filled with God. So He "is not only in Himself, where He was before He created the [finite] world, He is also everywhere" in the uncreated void (Grant 1981, 210). On this point—an uncreated void filled with God—Gassendi crossed the line between orthodoxy and heresy. His worldview, mixing, as it does, atomistic and Stoic components, would be especially influential in cosmological speculation right up to Newton at the beginning of the eighteenth century (Grant 1981, 119–213).

If Gassendi's cosmos can be called a Stoicized atomism, his ethics represents an important kind of Christianized Epicureanism. In the soulless universe of Epicurus and Lucretius, where everything is governed by chance and human beings are driven by desire and aversion, ethics is reduced to a hedonic calculation of pleasure over pain, and the pursuit of happiness is said to lie in the satisfaction of natural desire, which means observing limits and avoiding excess. For the ancient Epicureans, moreover, the gods have no input into this process; individuals are left to their own devices, without divine assistance, in determining how they ought to live. In Christian culture, this pagan teaching, which denied Providence and exalted pleasure, was often condemned as providing a license for vice and sin. But, in early modern Europe, discerning thinkers like Francesco Petrarch (1304–74), Ficino, Erasmus, and Montaigne argued that Epicurean doctrine had been misunderstood by Christians. The Epicureans actually taught that true pleasure lies not in unbridled sensuality but in self-control and moderation, not in the mindless pursuit of unquenchable lust but in leading a simple life devoted to satisfying basic needs for food and shelter as the best way of seeking contentment and repose. Those who sought to rehabilitate Epicurus's reputation argued that his teachings, far from being immoral, are essentially compatible with, and even conducive to, the cultivation of Christian virtue (Allen 1944).

Now, in the first half of the seventeenth century, Gassendi took a giant step in this rehabilitation. Not only was Epicurus (as modified by the Stoics) right about the fundamental structure of the universe, his moral philosophy, Gassendi insisted, provides the key to understanding how people should behave in order to achieve both the individual and the collective good. All human beings are driven, as Epicurus said, by desire and aversion, and the goal of human life is self-preservation. But Gassendi put a Christian spin on these pagan doctrines: In his hands, pleasure, pain, and self-preservation become the engines of divine Providence.

According to God's plan, human beings are full of natural needs for food, shelter, and procreation, needs that, if left unmet, will lead to misery and early death. But God is also an easy taskmaster who has attached a degree of pleasure to the satisfaction of these basic needs, and the more basic the need, the greater the pleasure in filling it. All people, then, are capable of obtaining the goods they seek. Indeed, they are so capable that in attempting to satisfy their appetites, they get in each other's way in their search for food and land and other goods, and the result is endemic strife. But, again, God is so wisely provident that He has endowed human beings with enough reason to see their way out of this impasse. They are led by the light of their natural reason to enter into a contract whereby they agree to hand over enough power to a political authority to maintain the civil peace so that each of them can be left free to get on with the pleasurable business of earning a livelihood. Once the state is created, Gassendi held, it must be obeyed, and there is no room for disobedience. But neither should the ruler be a tyrant because he would thus exceed the natural limits of his authority, whose purpose is to enable each subject to pursue self-interest, the satisfaction of personal needs, in such a way as to permit everyone else to do the same.

All, or most, "men" have enough reason to see the advisability of entering into and upholding the social contract that underwrites the collective peace and the pursuit of individual self-interest. But, according to Gassendi, only some "men" have enough reason to see that the true goal of human desire is self-preservation and that genuine self-interest, as Epicurus taught, lies not in the insatiable drive for more and more but in the repose and tranquillity that come from meeting one's natural needs. To go beyond this is a prescription for an excess of pain over pleasure. The simple life based on moderation is best. But only the wise few can see this, and so they are the only "men" who approach complete happiness (Sarasohn 1982; 1985).

Gassendi was a moral philosopher as well as a natural philosopher. Just as he attempted to provide new foundations for natural science, so he attempted to provide new foundations for Christian morals, and the common sources he drew on in both enterprises were the Stoic and Epicurean traditions. His combination of mitigated skepticism, Stoicized cosmology, and Christianized Epicureanism would have enormous influence in seventeenth-century England.

Descartes

René Descartes (1596–1650) was Gassendi's contemporary and, like Gassendi, he tried to establish new foundations for nothing less than all knowledge, both natural and moral. At La Flèche, one of the foremost Jesuit schools in France, he received a grounding in the traditional Scholastic curriculum, including Aristotelian science, which he later rejected. He then embarked on a military career. In his travels, he met and was influenced by Isaac Beeckman (1588–1637), a Dutch scientist who deepened Descartes's knowledge of mathematics, the only subject he singled out for praise from his previous education, and strengthened his conviction that the key to understanding nature lies in using mathematics to solve problems and arrive at scientific certainty. True to that conviction, Descartes went on to invent analytical geometry, which translates geometrical problems into algebraic form so that algebraic methods can be applied to their solution (Gaukroger 1995; Descartes 1968, 27–34).

Antiskepticism

Mersenne and Gassendi had argued that absolute certainty in the sciences is beyond the capacity of the human mind and that the best we can hope for from the sciences is probable certainty. But Descartes disagreed and set out to answer the skeptics by discovering the grounds for establishing necessary truth, conquering doubt, and building a new system to replace the false system of Aristotle. In this project, he proceeded by four stages, as laid out in his *Discourse on the Method of Properly Conducting One's Reason and of Seeking the Truth in the Sciences* (1637).

The first step was to divest himself of all the erroneous opinions he had acquired as a result of his prior education. As he said:

> I rooted out from my mind . . . all the errors which had introduced themselves into it hitherto. Not that, in so doing, I imitated the skeptics who doubt only for doubting's sake, and affect to be always undecided; for, on the contrary, my whole plan had for its aim assurance and the rejection of shifting ground and sand in order to find rock or clay. (1968, 50)

In this enterprise, he strikes a cautious, prudential note: "The mere resolve to divest oneself of all one's former opinions is not an example to be followed by everyone." With Galileo's recent fate in mind, he means to offer no challenge to public authority; in fact, he condemns "those meddling and restless spirits . . . forever imagining some reform of the State." Echoing Montaigne, Descartes says that his project is a private affair in which he hopes not to be followed by those who presume "to be cleverer than they are." But he is deeply ambivalent about the effect he wishes to have: "And if, my work having sufficiently satisfied me, I set it out here as a model for you, it is not on this account that I would advise anyone to copy it" (1968, 38). Later, as we shall see, he sings another tune and, in a Baconian vein, hopes for wide influence.

The second stage of Descartes's argument extends the first into new territory. Not only does he reject what he has been taught, but he proceeds along a course of systematic doubt—distrusting both his reasoning and the testimony of his senses, and even wondering whether he is awake or asleep. If asleep, he says, he might just be dreaming. But he is quick to recover from this deep uncertainty when he finds that, no matter how much he doubts, he cannot doubt that he is doubting and arrives at his first fundamental truth: "I think; therefore, I am" (Descartes 1968, 53). He exists and is essentially a thing that thinks, and this discovery leads him to the third stage in his quest for absolute certainty.

Third, as a thinking thing, he has innumerable thoughts, which may or may not have any truth, but Descartes has one more thought that carries its own conviction. He doubts and thereby recognizes his imperfection and, in that recognition, acknowledges the idea of perfection, which could only have been put into him by a perfect being. Hence God exists. We are now ready for step four. The existence of such a being guarantees that we can have true knowledge of a real world. A perfect God does not play tricks, and so we are not dreaming. The payoff is enormous. To the extent that we have "clear and distinct" ideas about the physical world, we can know nature, not through mere appearances, as Mersenne and Gassendi held, but for what it really is (Descartes 1968, 55–60).

Metaphysical Physics

What are these clear and distinct ideas, these real truths, that Descartes said we can have? First, each of us knows ourself as a thinking thing.

Second, we can also know objects outside our minds, including, first of all, our own bodies. God guarantees that this physical world is real and not illusory. But genuine knowledge of it does not come from the senses, because Descartes, like other contemporary philosophers, held that sensation is not an accurate representation of the outside world but only the product of a subjective encounter with it. Sensation yields a notoriously fallible account of the world. For true knowledge, Descartes relied, first of all, on mathematics, geometry in particular. We can know external objects clearly and distinctly to the extent that we can conceive of them as extended things reducible to their geometrical essences. This is the best knowledge—intuitive, innate, and purely intellectual. As Descartes said: "And now that I know God, I have the means of acquiring a perfect knowledge of an infinity of things . . . concerning physical nature, in so far as it serves as the object of the proofs of mathematicians" (1968, 149). Thus, Descartes, like Galileo, made a distinction between primary and secondary qualities of things. The primary qualities are their mathematically measurable characteristics—shape, size, position, and motion; the secondary qualities—for example, light, color, and texture—come to us through the unreliable senses and so cannot be trusted.

But, for Descartes, there was a second powerful source of true knowledge: metaphysics. In his *Meditations* (1641) and *Principles of Philosophy* (1644), he produced a philosophical physics deduced from first principles. His aim was nothing less than to replace the philosophical system of Aristotle and the Scholastics with his own new, and better, system, which would explain all the processes of nature, or at least provide the correct foundation for explaining them. The resulting Cartesian system was to have enormous influence on philosophy and science throughout the remainder of the seventeenth century. Here is an outline of its main features.

First of all, it is a mechanical philosophy. Everything in the physical world is made up of matter in motion and is put in motion initially by God. Second, Descartes put forward his own world-picture. The universe is "indefinite" in extent. Since Bruno's execution, the infinity of the universe had become a dangerous idea; Descartes let himself off the hook by leaving the question open. (A universe indefinite in extent may or may not be infinite.) The Cartesian version of a mechanical universe is not made up of atoms moving in a void, as Gassendi's is. Instead, the universe is full, a plenum in which there are no empty spaces and matter is not atomic in structure but infinitely divisible.

The only other thing in the created universe, besides bodies, is the soul, that is, the rational soul of each human being. (Animals do not have rational souls; therefore, they are mindless bodies, mechanical automatons.) Descartes is a dualist. All reality is made up of two substances, body and soul (or mind), radically unlike each other. The latter is thinking substance and unextended; the former, unthinking and extended. According to the terms of Cartesian metaphysics, there can be no immaterial extension and so no empty space. The point is fundamental for understanding the development of seventeenth-century natural philosophy. The physical universe is composed entirely of body, or bodies. Space does not exist independently of body, and so space without body is unthinkable (Koyré 1968, 99–109; Burtt 1954).

The Cartesian dualism is never made clearer than when Descartes proves his own existence:

> I thereby concluded that I was a substance, of which the whole essence or nature consists in thinking, and which, in order to exist, needs no place and depends on no material thing; so that this "I," that is to say, the mind, by which I am what I am, is entirely distinct from the body, . . . and moreover, that even if the body were not, it would not cease to be all that it is. (1968, 54)

This is a radical dualism. Human beings are composed of body and soul. But the two are "entirely distinct" from each other, and it is the soul that is defining.

On this basis, Descartes explained humankind's uniqueness. The rational soul is a spiritual substance, like God; but He is infinite, the human soul finite. Once created by Him, however, it is capable of surviving corporeal death, of life after death, of immortality. Descartes is making a metaphysical assumption in identifying this finite thinking substance with the immortal soul. He is insistent that human beings stand apart from and are superior to the rest of creation by virtue of their rational, immortal souls imparted by God. He is so insistent on this distinction because, he says, it is a subject "of the greatest importance: for . . . there is nothing which leads feeble minds more readily astray from the straight path of virtue than to imagine that the soul of animals is of the same nature as our own, and that, consequently, we have nothing to fear or to hope for after this life, any more than have flies or ants" (Descartes 1968, 76; Gaukroger 1995, 147–52, 184–6, 195–210).

The Cartesian dualism is designed, at least in part, to keep "feeble minds" on "the straight path of virtue" by refuting the idea that there is no significant difference between human and animal souls. The idea that beasts are on a par with people is dangerous, its results destructive of religion and morals. Were it to gain a hold on "feeble minds," they would suppose that, like "flies or ants," they are not immortal but live and die with the rest of creation. Then, without the carrot and stick furnished by belief in their immortality, they would lose the will to resist temptation, and their morals would be effectively subverted. The assumption in Descartes's reasoning is that this heretical attack on the Christian doctrine of personal immortality is indeed being made, that it is likely to be persuasive, and that it therefore needs to be answered. Descartes does not indicate where this attack is coming from, but he has in mind the same heretical sources that Mersenne has targeted. The naturalism of Pomponazzi and Vanini, and Bruno's pantheism, conflate the natural and the divine, as we have seen, and thus lend themselves to the subversive thinking that Descartes seeks to refute (Descartes 1968, 72–6). Pomponazzi's views were used to support the mortalistic heresy, the doctrine that the soul dies with the body, which had been condemned by the Lateran Council of the Catholic Church in 1513, but which continued to spread in radical circles throughout the sixteenth and early seventeenth centuries (Williams 1992, 63–71). In fact, Descartes claimed that, in answering the mortalists, he was carrying out the council's command to Christian philosophers "to refute their arguments and use all their powers to establish the truth" (Gaukroger 1995, 337).

But on theological questions, Descartes's metaphysics cut both ways. If the radical disjunction of matter and spirit could be used to defend the immortality of the human soul, it was thought by some critics to jeopardize the Catholic Church's teaching on transubstantiation, the priestly power during Mass to change ordinary bread and wine into the body and blood of Christ. If matter and spirit are so radically distinct and separate, how can the priest spiritually transform the material particles in question so that a miracle can be performed? Descartes's followers, later in the century, had to address the issue.

The Cartesian universe is filled with matter. How, then, does anything move if there is no empty space in which to move? Descartes argues that the universe consists of innumerable vortices made up of swirling masses of matter. Our solar system is one such vortex, centered on the sun and surrounded by other vortices, each with its own starry center. As there is no empty space within the vortex, so there is none between vortices. They are all crowded and rubbing against one another, like so many soap bubbles blown into a confined space (Fig. 11).

How, then, can the motion of all this moving matter be explained? Descartes was the first to arrive at a clear idea of the principle of inertia, an idea that lies at the roots of modern science, according to which, a body once in motion will move in a straight line at a constant speed indefinitely, unless acted on by an outside force. How, then, do all bodies keep swirling in their vortices around their centers? Given their inertia, all swirling bodies exert centrifugal force and, if unchecked, would leave their orbits and fly off in a straight line through space. But bodies are constrained to stay in their orbits because the matter in every vortex is exhibiting the same centrifugal force and exerting that force on every other vortex. The result is a balance of forces among and within vortices in which all bodies in a vortex—the planets in our solar system, for example—maintain their orbits forever.

The Cartesian vortices offered a powerful explanation of the structure of the post-Galilean, post-Keplerian universe. Though deeply flawed, it provided a truly mechanical account of the behavior of the stars and planets and replaced the old binary physics of Aristotle and the Scholastics with a new unified physics of moving matter. What it did not do was also clear enough: It did not provide an empirical, experimental account; it could not be confirmed by observations or by the mathematics in which Descartes put so much trust. His world-system rested on a theory that stated on a priori grounds what the universe must be and how it must work, but this theory failed to demonstrate, by some combination of numbers and experiments, what the universe is and how it works. That project would have to wait for Newton. Descartes's account, however, was a step in that direction because it provided a powerful (if not compelling) example to ponder over, argue about, and react against. Descartes, for his part, thought that it did much more than that (Westfall 1977, 30–9; Hall 1981, 107–23, 130–1).

Physics and Ethics

Descartes argued that his physics and cosmology laid the foundations for what he called "the study of wisdom," that is, "a perfect knowledge of all that man can know, no less for the conduct of his life than for the preservation of his health and the discovery of all the arts." All men can and should undertake this "search for wisdom"; in fact, it should be "their main preoccupation." The benefits would be both individual and collective: "Each nation is the more civilized and polished the better its members are versed in philosophy." The philosophy

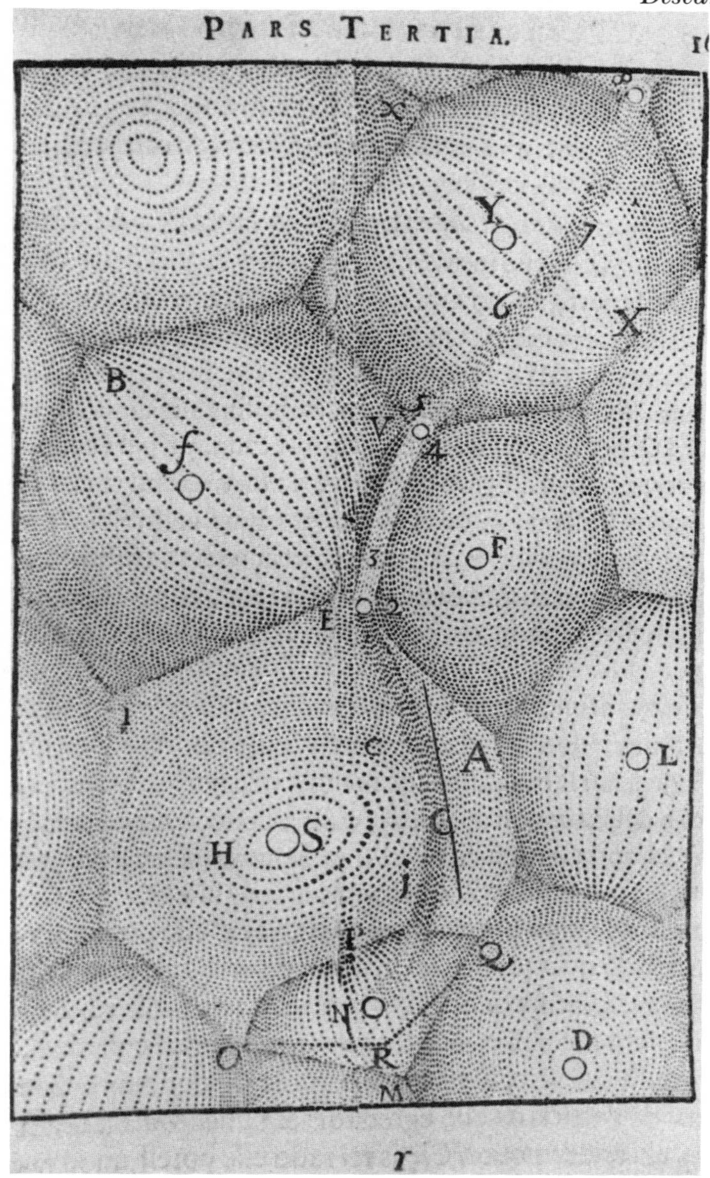

FIGURE 11 The Cartesian Vortices. The circle in the center of each vortex denotes a sun. (From René Descartes's *Principles of Philosophy*, 1644. Courtesy of Van Pelt Library, University of Pennsylvania.)

that is called for is, of course, Cartesian. On this basis, we might build "a practical philosophy" that would make us "masters and possessors of nature." On this basis, too, we might build an ethical philosophy that would make us masters of ourselves and provide the pathway to happiness (Descartes 1968, 173, 175, 174, 78, 47–9).

There is, moreover, an intimate connection between these two Cartesian projects—the scientific and the moral, the physics and the ethics. The bridge Descartes built between his science and his ethics was meant to be a two-way street. As explained below, the physics would convince men and women (he appealed to both) to lead moral lives, and the pursuit of virtue, in turn, would spur his followers to serve the public good, including the scientific project (Descartes 1968, 183, 186).

Cartesian ethics, like Montaigne's, are heavily influenced by Stoic and Epicurean teachings, and especially by Seneca, who combined both teachings in his moral philosophy. According to Descartes's reading of Seneca, "all men want to live happily but do not see clearly what makes a happy life." But Seneca has the answer, and Descartes agrees with him: "A happy life is one that is grounded in right and certain judgment," which provides the path to virtue and the regulation of "our desires and passions." Descartes then goes on to expand on the point by rehearsing the ethics first worked out in his *Discourse.* "It seems to me that each person can make himself content by himself without any external assistance" by using the tools already in his possession, his will power and his reason. The latter shows one the best course to be followed in any situation, and the former furnishes one with the resolve "to carry out whatever reason recommends." Such resolve, however, can go too far in denying pleasure and thus make virtue too "difficult to practice." But "the right use of reason" accommodates virtue to "licit pleasures" and makes it "easy to practice" (Cottingham et al. 1991, 256–62, 265; Levi 1964, 243–5, 281–3, 286–7, 290–5).

Up to this point, Descartes claims to be following Seneca and, like him, to be reconciling Stoic virtue and Epicurean pleasure. But even Senecan ethics is incomplete, he says, because it fails to teach us "all the principal truths whose knowledge is necessary to facilitate the practice of virtue and to regulate our desires and passions, and thus to enjoy natural happiness." Where Seneca falls short, however, Descartes's own metaphysical physics can be relied on to provide this necessary knowledge, "the truths most useful to us." According to Descartes, his physics contribute substantially to moral theory *and* practice. As to theory: "What little knowledge of physics I have tried to acquire has been a great help to me in establishing sure foundations in moral philosophy." As to practice: "These truths of physics are part of the foundations of the highest and most perfect morality" (Cottingham et al. 1991, 258, 265, 289, 368).

What are these compelling Cartesian physical truths? There are three: (1) the existence of a perfect God; (2) the immortality of our souls; and (3) the immensity of the universe. A knowledge of these three principles provides the foundation for both virtue and happiness. The first "teaches us to accept calmly all the things which happen to us as expressly sent by God." The second "so detaches our affections from the things of this world that we look upon whatever is in the power of fortune [including death] with nothing but scorn." The third principle teaches humility before the majesty of God's creation and, hence, before God himself, and so deepens our obedience to His will (Cottingham et al. 1991, 265–7, 289, 324–5; Koyré 1968, 99–109).

These principles provide the basis for a moral and emotional therapy. "If a man meditates on these things and understands them properly, he is filled with extreme joy." This joy itself is emotionally and physiologically beneficial. But more important is the effect of these principles on our desires and fears, especially the fear of death. The passions, left unregulated, are too insistent; the desires, in particular, promise more than they can deliver and, thus, disappoint and distract us from worthier goals. But the three Cartesian principles work to provide the necessary regulation. Assenting to their truth distances us enough from the sway of our passions to enable us "to pursue virtue—that is to say, to maintain a firm and constant will to bring about everything we judge to be the best, and to use all the power of our intellect in judging well" (Cottingham et al. 1991, 309–10, 296–7, 262).

Descartes has said that "each person" can undertake to lead a moral life "without any assistance." But he later recognizes what a difficult task he has set for all humankind: "It is true that the soul must be very detached from the traffic of the senses" if it is to assent to and act on his three moral principles. He claims, moreover, that we do not all have the same aptitude for success in the moral enterprise. There are two kinds of people in this regard—the more fortunate who can successfully steer their own course and the less fortunate who cannot and who may thus need some help from the former. Just as hunting dogs can be trained to obey their masters on command, he says, so human beings can be led to alter their emotional responses to sensory stimuli and exchange bad habits for good ones. So Descartes asserts that even "the feeblest souls" who get carried away by their "immediate passions" can be taught "to acquire a very absolute empire over all their feelings if enough skill is used in training and guiding them." In fact, everyone can benefit from such habituation techniques. It is not enough to know the three Cartesian principles; "practice is also required if one is to be always disposed to judge well." What is needed is "long and frequent meditation" on the truth of these principles so that it is "imprinted on our mind" and becomes "a settled disposition with us," which inclines us to act virtuously. Ever the optimist, Descartes claims that, with proper psychological training, all people might grow adept enough at regulating their passions to become independently virtuous and happy (Cottingham et al. 1991, 257, 310, 267; Descartes 1989, 47–9; Morgan 1994, 165–71, 185–211).

Science and Civilization

Descartes offers no concrete scheme for training people in this moral regimen, apart from self-help and the fact that the feeble-souled may need, and should receive, assistance from the sturdy-souled. But he does suggest what would happen if this regimen were to take effect. He claims that we would then be motivated less by selfish interest and more by the common good. Each would consider himself "a part of the community," delight "in doing good to everyone," and would "not hesitate even to risk his life in the service of others when the occasion demands" (quoted in Cottingham et al. 1991, 266).

Descartes singles out two kinds of such heroism. First, "every day we see examples of this love, even in persons of low condition [ordinary soldiers), who give their lives cheerfully for the good of their country." Descartes, himself a gentleman-soldier, pays respect to such heroic, albeit lowly, service and pays even greater respect to the cause of such sacrifice: "Sovereigns," he says, "are not [to be] overturned but strengthened" (quoted in Cottingham et al. 1991, 311; quoted in Keohane 1980, 211). A philosophical radical, he remains a social conservative. Descartes fears disorder, and science is meant to have the reverse effect, that is, to build order.

Besides military heroism, there is another kind, namely, scientific heroism, and Descartes regards himself as an example. He has published his natural philosophy in accordance with "the law which obliges us to procure, by as much as is in us, the general good of all men." He thinks that we can become the "masters and possessors of nature" and, in order to reach that goal, calls for a Baconian collaboration, a pooling of talents and experiments that would add up to a sustained adventure in scientific heroism:

> I would oblige all those who wish for the general well-being of men, that is to say, all those who are truly virtuous and not simply in appearance or who merely profess to be so, both to communicate to me the experiments they have already made and to help me to investigate those which remain to be done. (Descartes 1968, 81)

This, of course, is a peculiarly Cartesian version of Baconian collaboration because Descartes himself remains firmly in charge. In sum, the cultivation of private virtue can breed collective action for the common good and can do so, as Bacon also claimed, without disturbing public authority and the civil peace. In fact, for Descartes, echoing Bacon, science of the right sort raises the level of civilized life (Descartes 1968, 78, 81, 174–5).

But, of course, there is one big difference between the two thinkers. Bacon at least envisaged a scientific utopia, though how far he thought it might be realized is another question; Descartes never indulged in such utopian dreams. He remained wedded to the notion that everyone should obey duly constituted authority, and, in France, that meant submission to an increasingly powerful absolute monarch. Here is a difference not only between Bacon and Descartes but between English and French science, more generally, in the seventeenth century. As we shall see in the next chapter, many English natural philosophers, inspired in one way or another by the Baconian utopian impulse, addressed themselves to reforming both their science *and* their society; their French counterparts, with the spectacular exception of Gassendi, left the social question hanging. This political quietism may help to explain the victory of Cartesianism over Gassendism in late seventeenth-century France.

The Institutionalization of Science

Descartes stood at the beginning of another feature of modern science, the emergence of state-supported research organized around a national scientific society. The first of these societies was English, the Royal Society of London, inspired by Bacon and founded in 1660. But the French government established the Royal Academy of Sciences in Paris only six years later.

There were many precedents for government support of scientific research. The king of Denmark, for instance, had granted Tycho Brahe land and money for building a first-class observatory, and Kepler and Galileo had been given offices at court by their respective rulers, the Holy Roman emperor and the grand duke of Tuscany (Biagioli 1993; Kaufmann 1993; Westman 1980). The sixteenth-century Italian city-states, with their lively urban culture, were the settings for numerous court-related "academies," whose members met to discuss art, science, and literature and sought both intellectual and moral improvement. An early prototype was Ficino's Platonic Academy, founded by his patron Cosimo de' Medici (d. 1464), the ruler of Florence.

The earliest, strictly scientific academies seem to have been sixteenth-century Italian products, perhaps the first, the shadowy Secret Academy, being founded in Naples in the 1540s. These early Italian examples depended on a princely patron, used his libraries and research facilities, built his collections, and were, as a result, dedicated to serving him and not some larger public. Nature's secrets, it was felt, might be revealed to a courtly elite but lie "beyond the understanding of common people" (Eamon 1991, 44–5). In the next century, two Italian scientific societies stand out. The Academy of the Lynx-Eyed, of which Galileo and Della Porta were members, was founded in Rome in 1603 by Prince Federico Cesi (1585–1630) and died with him, and the Academy of Experiment met in Florence between 1657 and 1667 at the behest of the reigning duke of Tuscany. In these two later societies, a shift away from the courtly ideal of privileged science to a courtly ideal of public science, which echoed Bacon, can be detected. In 1616, for instance, Cesi, perhaps under Galileo's influence, rededicated his academy to "the propagation of knowledge, to the communication and advancement for public utility of the virtuous toil and the results made by them [the Academicians]" (Eamon 1991, 46–7). Even so, these Italian societies were short-lived, locally based, socially exclusive, and closely tied to their aristocratic patrons, unlike the more impersonal and enduring national organizations created later on in France and England (Biagioli 1993; Eamon 1991; Hall 1962, 238–47; Redondi 1987, 68–106; Yates 1983).

In France, during the middle third of the seventeenth century, a number of groups had met to discuss science both in Paris and the provinces. Mersenne had been particularly active in bringing scientists together and sustaining an informal but vast network of scientific correspondence. We have already seen him calling for academies for policing the alchemists. Cardinal Richelieu (1585–1642), the king's chief minister, had given official sanction to scientific research and communication. But not until Jean-Baptiste Colbert (1619–83) took charge in the reign of Louis XIV (1643–1715) did the new science become a matter of government policy. Colbert believed that every aspect of French intellectual and artistic life should be brought under state supervision, and one result was the creation of the Royal Academy of Sciences (Fig. 12). It functioned from then on as a select society of experimental philosophers, chosen for their talent and not their social background. It was limited to twenty or thirty members, including such foreign notables as the Dutch Cartesian Christiaan Huygens (1629–95) and the Italian Gian-Domenico Cassini (1625–1712), both lured to Paris by the promise of a handsome pension and proper research facilities paid for by the state. In return, the academy was expected to engage in collaborative research directed toward providing solutions to practical problems, such as how to determine longitude at sea. Such knowledge was meant to be useful to the kingdom, broadly conceived, including the military, the merchants, and the maritime community (Brockliss 1992, 55–63; Brown 1934).

The new science spread by other means as well. The first scientific journals began to be published in the late seventeenth century, notably, in France, the *Journal des Scavans* (*Journal of the Learnèd*), the first of its kind, in 1665. Both books and journals were published in the vernacular as well as in Latin. Descartes was a master of polished French, as Galileo had been of Italian, and thus made his version of the mechanical philosophy accessible to a large number of literate men and women. In the opening paragraph of the *Discourse*, he insisted, furthermore, that reason "is naturally equal in all men" and so, despite his backpedaling on the issue, effectively dissolved the barriers, both professional and clerical, to philosophical thinking and invited everyone to join in the search for truth. Descartes, as we have seen, was no democrat, either social or intellectual. But this rhetorical leveling of the intellect represents a striking departure from the elitism, illuminism, and social exclusiveness of so much earlier science (Descartes 1968, 27, 38–9; Brockliss 1992, 63–8, 70).

The Spread of Cartesianism

His message took, thanks, in large part, to the efforts of his followers, like Claude Clerselier (1614–84) and Jacques Rohault (1618–72), who, in the three decades after Descartes's death, saw to it that his books

FIGURE 12 A Royal Library in the Reign of Louis XIV. Plans for military fortifications are being discussed by the figures on the right. (From the *Histoire* of the reign by Monsieur de la Hode, 1741. Courtesy of Van Pelt Library, University of Pennsylvania.)

were printed and reprinted and themselves lectured and wrote textbooks on Cartesianism. Clerselier even exhumed Descartes's body in Sweden, where he had died, brought it back to Paris, and gave him an unofficial state funeral. In 1686, Bernard de Fontenelle (1657–1757) published his *Conversations on the Plurality of Worlds,* which aimed at a female readership and proved the most successful attempt to popularize Cartesianism. It had gone through seven editions by 1714. Descartes continued to enjoy a significant female following perhaps because, as one of his early followers remarked, his assertion of the principle of intellectual equality implied that "the mind has no sex," that the gender barrier to scientific thinking might also be broken down. Cartesianism, in some quarters, did not go unmixed with intellectual and social radicalism (Harth 1992, 8–10, 81–2).

But its growing success depended, of course, on other factors. In France, by the end of the seventeenth century, the mechanical philosophy had triumphed over its two main rivals, Aristotelianism and Paracelsianism, and the version of mechanical philosophy that had succeeded was Cartesian rather than Gassendist. Initially, at midcentury, Gassendi and Descartes were both theologically suspect, the former for his revived Epicureanism and the latter for his radical disjunction of matter and spirit, which seemed to undercut any rational explanation for transubstantiation and thus to open the door to Protestantism. But later Cartesians repackaged their master's teachings to make them more palatable to the church. Unlike dogmatic Cartesians, they played down metaphysics and soft-pedaled the issue of the separation of matter and spirit. Careful distinction, moreover, was drawn between inert and living matter, and animals were not denied souls.

In addition, the church in France did not have the power to enforce a rigid Aristotelian orthodoxy, as it did in Italy. There was no Inquisition, and censorship was within the jurisdiction of the civil courts. Thus, philosophers of all sorts had space to breathe, and, in the resulting competition of ideas, the Cartesians won the day by making a better case for their views than their rivals did. In France, their victory was so complete that it was not until the 1740s that Cartesianism lost its grip in favor of Newtonian natural philosophy and vortex theory gave way to the law of universal attraction as a superior way of explaining the structure and behavior of the heavens (Brockliss 1992, 68–82).

Science in Seventeenth-Century England

THE SCIENTIFIC REVOLUTION culminated in England. The experimentalism initiated by Gilbert, Bacon, and Galileo, the mitigated skepticism of Mersenne and Gassendi, the mechanical hypothesis of Mersenne and Descartes, the creation of a government-sponsored scientific society for organizing and stimulating scientific research, and, finally, a compelling answer to the question of what keeps the planets in their orbits around the sun—all these developments were brought to fruition in seventeenth-century England. The story of English science is made more interesting because it takes place in the context of the English Revolution, a long period of religious and political crisis that stretched from the late 1630s to the Revolution of 1688–9. During this half-century, England was finally transformed from being a hereditary monarchy, with the king in charge, into a constitutional government in which the king's power was limited by the laws of parliament. The shape that science took in England was affected at certain points by this revolutionary upheaval. We shall pursue this story from Bacon's first followers, the early Baconians, in the 1630s and 1640s through to the achievements of Isaac Newton in the 1680s. But, first, the setting in which the new scientific thinking occurred must be briefly sketched in.

The English Revolution

By the 1630s, England was a deeply divided kingdom. In politics, the division was between the king, Charles I (1625–49), and the growing opposition to him in parliament. Charles asserted the royal prerogative—

that is, his power to make decisions and, especially, to raise taxes—on his own authority without having to obtain the consent of parliament. His opponents had other ideas and insisted that parliament must share power with the king over taxes, if not other matters as well. In religion, the basic split was between the king, who wished to govern the state church from the top down through bishops appointed by himself, and the Puritans. The latter insisted on greater lay participation in the government of the church at the expense of the bishops' power and a progressive modification, or "purification," of church ritual and doctrine in the direction of Calvinism. According to the Puritans, sinners are saved by faith, not by works, and church services should emphasize the preaching of God's word rather than the performing of elaborate ceremonies. On all three issues of doctrine, ritual, and government, the state church strayed too far toward what the Puritans derisively called "popery" and needed to be reformed.

By 1641, the king had been forced to call parliament because he needed to raise taxes to build an army. Many expected that a religious and political settlement could now be worked out. Hopes ran high, driven by the widespread belief that the millennium was fast approaching. As the apocalyptic books of the Bible foretold, Christ or his appointed ones would appear and usher in a new order to last a thousand years. In England, good government in church and state would be restored, and the age-old problems of poverty, hunger, and disease would be relieved, if not eliminated. Throughout the world, the sainted Protestants would win out over the diabolical forces of the pope.

Among these millenarians were many scientific thinkers and artisans, followers of Francis Bacon, who, in many cases, were also Puritans and friends to parliament. The chief organizer of these Baconians and their effective leader was Samuel Hartlib (d. 1662). For three decades, from the 1630s through the 1650s, he devoted himself to the millenarian cause, "the reformation of the world," as he called it, and to deploying Baconian science as a key tool in achieving such a goal. From Bacon, Hartlib and the others took the idea that science should be empirical, experimental, and practical, dedicated to relieving misery and improving the human lot. They agreed with Bacon that the state should be responsible for funding and directing scientific activity. Everyone should be put to work and production and prosperity maximized. The result would be a Baconian utopia—New Atlantis realized and the millennium achieved (Boyle 1966, 6:113; Trevor-Roper 1967; Tuveson 1949, 71–112; Webster 1975, 1–31).

One of Hartlib's protégés, Gabriel Plattes (ca. 1600–44), wrote *A Description of the Famous Kingdome of Macaria,* which Hartlib published in 1641 and which comes closer, perhaps than anything else, to providing a blueprint for Hartlib's Baconian vision. Under the king and parliament, according to this tract, councils of husbandry, fishing, and trade should be established. Their task would be to maximize production and wealth, "whereby the Kingdom may maintain double the number of people which it doth now, and in more plenty and prosperity than now they enjoy." In addition, there should be a government-sponsored college or Society of Experimenters modeled on Bacon's Salomon's House, which would be devoted to increasing "the health and wealth of men." To achieve their ambitious goals, the economic councils should have enormous power. The Council of Husbandry, for example, would impose a 5 percent death duty on "every man's goods" for the improvement of trade and farming. All landowners, for their part, would be expected to improve their lands "to the utmost," and anyone who failed to do so would pay a penalty to be "yearly doubled, till his lands be forfeited, and he banished out of the Kingdom, as an enemy to the commonwealth." Religion would also be carefully regulated, religious opinions being vetted by the state before being published, so that conformity might be encouraged and the civil peace preserved (Plattes 1641; Webster 1979; Davis 1981, 85–103, 312–24).

The widely expected millennium, of course, never occurred; as one date would pass, another would be set a few years ahead. What did happen was nonetheless revolutionary enough. The king and parliament could not agree in 1641 and the next year took to the battlefield to decide their differences. After six years of bloody civil war, parliament defeated the king. In the aftermath, Charles I was tried and executed for treason on 30 January 1649. The ancient monarchy was destroyed and a kingless republic proclaimed. The next decade, the 1650s, proved to be a period of breathless experiments in politics and religion. Successive republican governments were set up only to fail, each in turn, because none could build enough support in a deeply divided nation. In religion, the old state church was destroyed, and the censorship of pulpit and print collapsed.

In this vacuum of control, radical opinions, long forbidden, came out in the open, and the radical sects (small groups of like-minded religious believers) flourished as never before or since. Their views often derived from the Radical Reformation of the previous century (26–27). The very names of these sects are evocative. The Levellers called for universal manhood suffrage and the elimination of the property qualification to vote. The Independents attacked the very idea of a state church and the taxes collected to support it. The Seekers asserted the capacity of every person, man or woman, to find his or her own way to God. The Ranters called for absolute freedom of expression. The Quakers exalted the divine light within each person that equalizes us all and used it to attack every kind of hierarchy. Perhaps most radical of all were the Diggers, whose leader, Gerrard Winstanley (ca. 1609–after 1660), asserted that God and nature are one and argued for, and took steps to effect, a redistribution of land so that it could be worked communally for the common good. These various sects preached and practiced what they preached. In 1656, the Quaker James Naylor (ca. 1617–60) entered Bristol on a donkey, with women scattering palm branches in front of him as witness to his belief that Christ's perfection could be realized within us all. A few years earlier, the Diggers actually set up communal farms by going out and "digging up" land that did not belong to them, but which they claimed should be available for the use of the poor and needy (Hill 1972; Williams 1992).

There was widespread hostility to such radicalism, which was called "enthusiasm" by its enemies. This pejorative referred to the claim made by many sectaries that they were divinely inspired, a claim that justified their attacks on the old order but was so much empty humbug to their opponents. The fears of the propertied about imminent social revolution reached a state of near hysteria in the early 1650s. Opinion divided and hardened between, on the one hand, the sects, often representing the poor and disadvantaged, who sought "to turn the world upside down," and, on the other, all those, especially the gentry and the propertied, who sought to turn it right side up again. Landowners would not tolerate a threat to property rights and social hierarchy and eventually won out over their enemies. The radical sectaries were defeated and silenced. The hereditary monarchy, the hereditary House of Lords, and the episcopal state church were all brought back to power in the so-called Restoration of 1660, which put Charles II (1660–85), Charles I's son, on the throne (Coward 1992, 210; Thirsk 1992, 184).

In this context of revolutionary political and religious turmoil, natural philosophy also developed in some striking new directions. In addition to the Hartlib group, there are two more developments that must be singled out for special mention: the philosophy of Thomas Hobbes and the revival of Paracelsianism.

Hobbes and Hobbism

In the annals of political philosophy, no one ranks higher than Hobbes (1588–1679). His natural philosophy is also important, and what is central to our story is the close connection among his scientific, political, and religious thought. In particular, he argued that the messy, fractious world of politics could be made subject to scientific understanding and control and that he, in fact, had arrived at such an understanding in his greatest book, *Leviathan* (1651). Its political truths, Hobbes argued, are as self-evident as those of the most demonstrable science of all—geometry. Furthermore, he claimed that were his scientific rules of politics to be taught in the universities to the sons of the gentry and then imparted by them to the people, the best possible state might be established, that is, the only kind of state that could produce civil peace and put an end to chronic divisiveness and turmoil in England. Hobbes thus intended *Leviathan,* his great work of philosophy, to be a prescription for healing his country's wounds and building a new and lasting order (Johnston 1986).

Materialism and the Human Predicament

What was needed first was to start from correct premises. For Hobbes, this meant materialism: the world is made up of nothing but material particles in motion. What was the philosophical basis of Hobbes's materialism? Everything in the world, he claimed, is constantly changing and therefore moving. But nothing in the world can move itself, he asserted, and nothing can be moved but bodies in space, and then only by other bodies all the way back to the first cause or God, the supernatural Creator of the natural world. Hence all reality is material (Hobbes 1968, 401–2; Tuck 1993, 299–301).

Like Mersenne and Gassendi, who were Hobbes's friends, he was also a mechanist: Science is the study of the universe conceived as a vast machine made up of moving atoms. But, unlike the other two (and Descartes as well), Hobbes made no distinction between body and soul. He was a thoroughgoing materialist: All reality is material, and there is no empty space; it is full of material particles. There is, moreover, no separate spiritual realm; souls or spirits, if there are any, are merely more subtle or attenuated configurations of matter. What, then, did Hobbes make of God? Hedging on the question, he asserted God's existence but denied that we can say much else about Him because finite creatures cannot pretend to fathom the infinite Creator of the universe. On this point, Hobbes could not have departed further from his French associates. In his view, those who claim to know God or to speak for Him—the Christian clergy or those putatively inspired by God—are either self-deluded or deliberately set out to deceive "the simple people" in order to gain power over them. To make matters worse, Scholastic philosophy, taught in the universities, insists on the existence of "incorporeal spirits"; for Hobbes, the materialist, this is a contradiction in terms, which leaves the people more bamboozled (Hobbes 1968, 93, 168, 171). In *Leviathan,* he pours scorn on the crafty clergy and schoolmen and on the gullible public who are taken in. His book, in turn, was reviled and censored as a byword for atheism, and a clever, persuasive one at that. And there was worse (Tuck 1993, 319–45).

According to Hobbes, human beings are, like everything else, made up of matter in motion. Mental functions, like bodily ones, are purely physical, and it is Hobbes's materialist theory of mind that is especially important. It offers a mechanical push/pull psychology. We are driven by two kinds of material motions inside ourselves, namely, our appetites and aversions. We are pulled toward what *appears* attractive and pleasurable to us and pushed away from what *appears* unattractive and painful. Perception is thus crucial in Hobbes's thought. Like his contemporaries Mersenne, Gassendi, and Descartes, Hobbes was a radical skeptic as to the validity of sense experience. Through the senses, we do not know the world as it really is but only our subjective perceptions of it. The three French philosophers, however, relied on God to extract us from this skeptical predicament. We might not have direct access to the real world, they said, but God will insure that our mental and perceptual functions are such that science can still be done and probable certainty (and, in Descartes's case, even rationally deduced truth) can be reached. But Hobbes did not take this way out. God for Hobbes, we recall, is an unknown quantity, a hidden God. We cannot say what He does or does not do for us. Hobbes leaves each of us to our own epistemological devices, trapped in our own perceptions without sensory access to anything beyond ourselves. This is psychological individualism, solipsism, and isolation with a vengeance. And our situation is even bleaker than this (Hobbes 1968, 85–99, 118–30, 134–9; Tuck 1993, 279–303).

Human desires are insatiable and, if left unchecked, produce unending destructive conflict, the war of all against all. Hobbes gives three reasons for this: "First, Competition; Secondly, Diffidence; Thirdly, Glory." Concerning the first, he says that men often covet the same thing, for example, land, and that when they do, they will go so far as to fight and die to get it. He says, of the third reason, that human vanity also leads to conflict, when, as all men do, each one thinks himself superior to his fellows and expects them to acknowledge it. The second of Hobbes's three reasons for civil war is perhaps the most important. Each of us, locked in a world of our own cravings triggered by our perceptions, does not and cannot know what our competitors are going to do. So "there is no way for any man to secure himself, so reasonable, as Anticipation." This is the psychology of the preemptive strike. Even if one does not want more than he has, he is forced to grab whatever he can in order to protect what he already has from what he *thinks* his competitors might do. The best defense is offense. The upshot is that, without effective government, things may degenerate into the state of nature, where life is "solitary, poore, nasty, brutish, and short" (Hobbes 1968, 160–2, 183–8; Tuck 1993, 304–14).

Leviathan

For Hobbes, the only way out of this impasse is for people to enter into a contract to submit to a sovereign with enough power to enforce obedience and its fruit, which is lasting civil peace (Fig. 13). Among other things, the sovereign must have the power to take his subjects' property whenever public need requires. For example, according to Hobbes, excessive private wealth makes for sedition; that is, rich men who are also ambitious (for Hobbes, not an unlikely combination) will use their wealth as a lever for their ambition "because all things obey money," and thus challenge the sovereign's authority. In such cases, he must step in and make sure "that the commonweal receive no prejudice." Either he may confiscate some of these private riches, thereby "diminishing their heaps," or, in extreme cases, a covetous man may be "cast out of Society, as cumbersome thereunto." Banishment and confiscation, we recall, constitute the very punishments that Plattes and Hartlib would impose on landlords who failed to improve their lands "to the utmost." Now it is not they but greedy, ambitious disturbers of the public peace who must pay such draconian penalties (Letwin 1972, 160; Hobbes 1968, 223–8, 297, 209).

Hobbes was a dogmatist and an absolutist. The sovereign's power, if peace and order are to be kept, must extend not only to land and wealth but also to religion and education, and even to scientific research. The sovereign must prescribe the doctrines and forms of public worship so that there are no competing authorities in the state who might differ

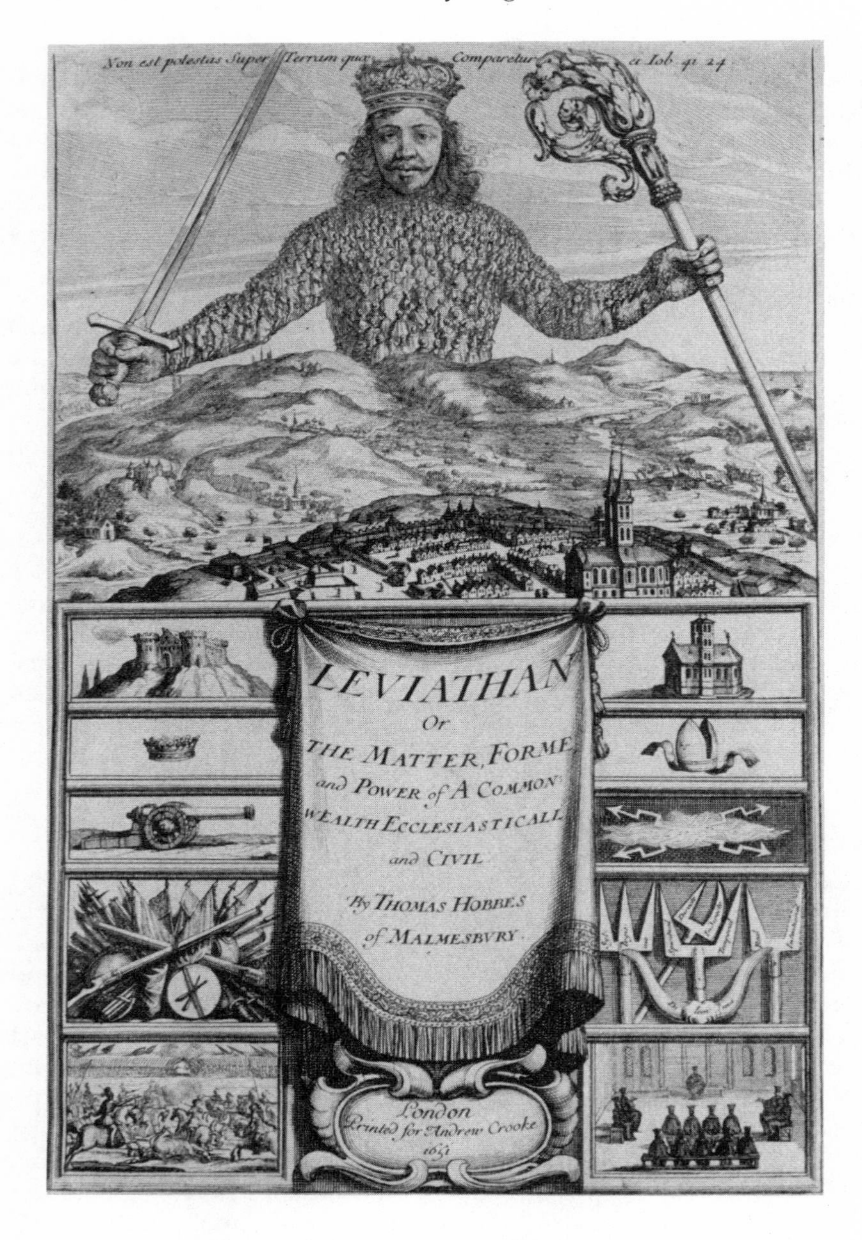

FIGURE 13 The Frontispiece to Thomas Hobbes's *Leviathan* (1651). The king is represented at the top. Notice that his subjects make up his arms and torso, but *not* his head. This can be read as a striking visual metaphor for Hobbes's theory that to escape the horrors of civil war, every individual must subject his private will to the will of the sovereign, who thus becomes, as Hobbes says, "that Mortal God, to which we owe . . . our peace and defence." (Hobbes 1968, 127) (Courtesy of Van Pelt Library, University of Pennsylvania.)

with and resist the sovereign himself. He must also take particular pains to indoctrinate his subjects in their duty to obey his commands. As Hobbes realized, naked power and harsh punishments are not enough, by themselves, to keep the peace. The people must also be taught to see why they must obey, namely, that the alternative is too dreadful to contemplate, that given people's atavistic proclivities, disobedience will inevitably lead to the war of all against all (Hobbes 1968, 377; Tuck 1993, 319–35).

Hobbes's obsession with the need for thought control as a guarantee of unity and peace even extended to science. He was deeply suspicious of a scientific society, like the Royal Society of London, founded in 1660, where independent inquiry, unsupervised by the sovereign, might be conducted. Hobbes thought such a think tank might function as a new priesthood, as an alternative and rival authority in a state, which might gather support and challenge the sovereign for control. The result would be more civil war. So the stipulation and interpretation of belief, and even the conduct of science, must be left firmly in the hands of the civil sovereign and his deputies. To solve the problem, Hobbes recommended that much of the Scholastic curriculum, so conducive to wrangling and controversy, be discarded and *Leviathan* put in its place as required reading in the universities. What better way to jump-start the peace process, he thought, than to instruct the children of the gentry in his perfect science of politics, carrying, as it did for him, an all but mathematical level of conviction (Hobbes 1968, 727–8; Johnston 1986; Shapin and Schaffer, 1985)?

Radical Paracelsianism

Like the new Hobbism, the Paracelsian revival of the 1640s and 1650s is important for understanding the intellectual context of English science. For a century, the ideas of Paracelsus and his followers had spread in England and affected the practice of alchemy and chemical medicine (the treatment of disease with medications rather than by Galenic methods). The outcome was a kind of compromise—physicians might use either Galenic or Paracelsian therapies or, for that matter, both. But, beginning in the 1640s, Paracelsianism developed a new ideological edge reminiscent of its earlier career and of Paracelsus's own thinking.

There were several reasons for this. First, the revolt against the monarchy saw an attack on royal monopolies in law, manufacturing, and trade. There was also a medical monopoly incorporated under the crown, namely, the Royal College of Physicians, which excluded surgeons and apothecaries, who were likely to be Paracelsians. In the heady atmosphere of the 1640s and 1650s, the champions of the excluded attacked the college as outdated and corrupt and called for a new, democratic system of public health inspired by Hermetic natural magic, Paracelsian alchemy, and chemical remedies. Second, during the same period, the writings of Paracelsus and his followers were translated and published in large numbers. Hartlib himself was at the center of much of this activity. Third, Paracelsianism, as we have seen, was a millenarian movement from the start, and the idea of a Paracelsian chemical millennium took a new lease on life in mid–seventeenth-century England. Many of the men in the Hartlib circle, like Frederick Clodius (fl. 1640–60), Robert Child (fl. 1640–60), and Cressy Dymock (fl. 1640–60), shared this ideal and looked forward to "a new heaven and a new earth," when all secrets of nature would be revealed and a new world order of peace and prosperity established by chemical means. The Hartlibians were Baconians *and* Paracelsians, and the two impulses did not always sit well together, as we recall Bacon's hostility to Paracelsian pretension and magic (Trevor-Roper 1985, 178–97). Finally, there was a link between Paracelsianism and the sects: They both appealed over the heads of earthly authorities to divine inspiration as a source of knowledge and power, a source not open to outside verification. One simply had to take the saint's word for it (Rattansi 1963).

The radical sectary John Webster (1610–82) represents an important example of this Paracelsian-sectarian alliance. In *The Examination of Academies* (1654), he launched a sweeping attack on the traditional curriculum, based on Aristotle and Galen, served up at Oxford and Cambridge. Webster went on to call for new science instruction built around Bacon and Fludd. But, as Seth Ward (1617–89) and John Wilkins (1614–72), both influential Oxford dons, pointed out in their reply defending the universities, *Vindiciae Academiarum* (*The Defense of the Academy*), also published in 1654, "there are not two ways in the whole world more opposite than those of the Lord Verulam [Bacon] and Dr Fludd, the one founded upon experiment, the other upon mystical ideal reasons." This reply was a watershed in English scientific development, comparable to the attack launched on Fludd and the magicians by Mersenne and Gassendi in France a few decades before. An increasingly conservative Baconianism, represented by Ward and Wilkins, was distancing itself from radical Paracelsian influences; it was the former that would go forward as the mainstream scientific movement, while the latter slipped into the shadows from the late 1650s on. With the Paracelsian radicals was also jettisoned much, but not all, of the social idealism that the Hartlib circle had represented. This process will receive close attention in the next section (Trevor-Roper 1985, 191; Debus 1970; Rattansi 1963; Hill 1972, 231–46; Jacob 1978).

A Watershed in English Science

Wilkins and the Latitudinarians

By 1654, the man of the hour in English science was Wilkins, the warden of Wadham College, Oxford, and the university's defender against attacks by radicals like Webster and Hobbes. Ordained a priest many years before, he had strong Puritan and parliamentary connections and was married to the sister of Oliver Cromwell (1599–1658), the Puritan leader of the victorious parliamentary army and the most powerful politician in the new republic set up after the regicide in 1649. Through a combination of his talents and connections, Wilkins was appointed to his Oxford post in 1648 and then became master of Trinity College, Cambridge, in 1659. After the Restoration, despite his Cromwellian connections, he rose to become a bishop. Though Wilkins was an important preacher and cleric, science was always one of his main interests. He popularized the new astronomy of Copernicus and Galileo and was at the center of the scientific circles in London and Oxford that led to the founding, in 1660, of the Royal Society, the first national scientific organization in the world.

Among these London circles was the one based at Gresham College, set up in 1597 and funded by a bequest from the wealthy London merchant Sir Thomas Gresham (1519–79). The college was established at Gresham's house, where seven Gresham professors were meant to live and lecture. From the start, its professors of geometry and astronomy made an important, though often overlooked, contribution to the development of English science, a contribution that Wilkins acknowledged and extended. These Gresham men—Henry Briggs (1561–1630), Edmund Gunter (1581–1626), and Henry Gellibrand (1597–1636), for example—built on a native tradition of applied mathematics aimed at serving the needs of merchants and the state. They invented and perfected mathematical instruments and the techniques for using them. These could then be applied to solving practical problems related to navigation, surveying, and industrial design, like finding latitude, drawing a map, or making guns and clocks.

It has been argued that the efforts of the Gresham men helped to create an atmosphere favorable to the new science. First, their mathematical investigations, directed to solving practical problems, may have contributed to Baconian experimentalism, that is, to the idea that scientific knowledge comes not from speculation but from induction and controlled testing. Second, the mechanical instruments they devised, and the successes they scored in deploying them, may have contributed to the abandonment of Scholasticism and the acceptance of the mechanical philosophy, to the notion that Aristotle's qualitative physics is unhelpful and that nature is better conceived on the basis of a quantitative physics of matter and motion, such as Galileo and others were then working out. Certainly Wilkins, in praising the mathematical achievements of the Gresham professors, argued that natural philosophers would do well to go to "tradesmen's shops" and be instructed by hands-on mathematicians because, as he said, there is "a divine power and wisdom . . . even in those common arts." The practical, mechanical thinking of the Greshamites seems to have fostered the emergence of the mechanical philosophy in England (Bennett 1986).

Wilkins himself was committed to an experimental and mechanical philosophy heavily influenced by Bacon and Mersenne. In fact, his writings have been called "a popular version of Mersenne." According to Wilkins, only God can know things as they really are; we must rely on our fallible senses. But this is enough to allow science to be done. If we conceive of nature in mechanical terms, make careful observations, and frame experiments that yield mathematically measurable results, it should be possible to arrive at probable certainties about phenomena—not absolute truth, to be sure, but usable, communicable knowledge nonetheless. Wilkins rejected the claims of all those, like Fludd, the Rosicrucians, and the radical Paracelsians, who asserted that their knowledge comes from divine inspiration. "This was the mystical way, repeatedly rejected by Mersenne [and Wilkins] as nonsense; only God can know the essences of things," and claims to unique, secret, unverifiable knowledge should not be accepted (Aarsleff 1976).

Echoing Bacon, Wilkins also insisted that science should have practical application, as, for example, in the invention and use of labor-saving machinery. He interpreted these technological benefits of science in theological terms also reminiscent of Bacon, as being "so many essays, whereby men do naturally attempt to restore themselves from the curse upon their labors" that God pronounced at the Fall as punishment for original sin. Here, Wilkins was giving new life to Bacon's idea of the Fortunate Fall, which he may, in turn, have derived from Bruno and Paracelsus. The idea was fundamental to Wilkins's natural religion and that of the group he led in the restored Church of England known as the latitudinarians (Aarsleff 1976, 363; Rivers 1991, 31).

According to this group, human beings are not utterly depraved creatures who must depend entirely on God's mercy for salvation, the view shared by strict Puritans, including many radical sectaries. On the contrary, the latitudinarians held that humankind has a natural capacity for goodness and the free will to act upon its generous impulses. These gifts were tarnished and diminished by the Fall but not obliterated and are always available to be nurtured and acted on. We are essentially sociable creatures, as Aristotle and the Stoics (especially Seneca and Cicero) have taught, capable of working together for our mutual benefit. Public and private good can be obtained at the same time through the same course of action. Science is one important avenue for such beneficial cooperation; the discovery and spread of scientific knowledge promote trade and industry, Protestant unity, peace and prosperity, the welfare of each and all (Shapiro 1968; Rivers 1991, 58–60, 73, 77–88).

Their Moral Theology

Many English thinkers shared this rosy picture of human capacity and the role of science. It was a picture that represented a persuasive answer to both Puritanism and Hobbism. Hobbes's materialism called all religion into question and offered a particularly pessimistic view of humankind as ruthlessly acquisitive. Moreover, Hobbes's cure, ruthless statism, was perhaps worse than the disease. Extreme Puritanism also veered between two extremes—on the one hand, all human beings depraved and incapacitated by the Fall from doing good and, on the other, at least some people, the saints, miraculously saved and inspired by the gift of God's grace and thus possessed of extraordinary power and knowledge, such as sectaries, Paracelsians, and Rosicrucians claimed. Wilkins's answer to Puritans and Hobbists alike endowed human beings with the wherewithal from birth to be responsible, with *help* from God, for their salvation (*pace* the extreme Calvinists) and also to be responsible, along the way, for using their intellect to do science and improve the human lot without the need of Leviathan (*pace* Hobbes) (Rivers 1991, 34–40, 44–5, 59–60, 63).

Indeed, Wilkins's response, and that of his latitudinarian colleagues, to Hobbes and others, an answer worked out at Oxford during the

1650s and developed after the Restoration, was more specific and robust than has been seen by historians. The result was a moral theology and social ideology in which Baconian science played a key role.

Hobbes, Hartlib, and Plattes, we recall, had argued that the state should have the power to impose ruinous taxes, confiscate the property of its subjects, and even banish its former owners out of the commonwealth. The purpose of such forced redistributions of wealth would be either to keep the civil peace by cutting overmighty and fractious subjects down to size (Hobbes) or to maximize production by transferring land from the least efficient to the most efficient farmers (Hartlib and Plattes). But, according to the Wilkins group, individuals can build a peaceful, prosperous, and just society without the need to resort to such draconian methods, and the latitudinarians thus distanced themselves from Hobbist and Plattist statism and absolutism just as they did from Paracelsian and Rosicrucian magic and enthusiasm.

According to Wilkins and his fellow latitudinarians, we are endowed by nature with enough reason to allow us, with God's help, to earn our own happiness, both in this life and hereafter, and, in the process, to increase the happiness of our earthly community. Our task, then, is to use our reason to cultivate the virtues that will provide this reward. The practice of virtue requires effort and discipline, but it leads to pleasure and happiness. Essential to the pursuit of virtue are religion and science—the study of God's word, which is Scripture, and the study of God's work, which is nature. Religion and science together will nurture the virtues and excite their exercise in honest work, charity toward others, and the discovery of natural knowledge.

Their Social Ideology

The result of such a disciplined and virtuous life will be great riches. But such wealth is not to be conceived, in bleak Hobbist terms, as the product of "a perpetuall and restlesse desire of Power after Power, that ceaseth only in Death" (1968, 161). Rather, wealth, as conceived by Wilkins and his followers, is something entirely different. Echoing a sixteenth-century humanist tradition reaching back to Erasmus, Wilkins, for instance, wrote of the wholesome pursuit of "sufficiency" as opposed to the unhealthy Hobbist drive for unlimited accumulation (Todd 1987, 118–75). By sufficiency, Wilkins meant that amount of wealth it takes for us to satisfy "the occasions" associated with the rung each of us occupies on the status ladder. Sufficiency, then, varies with social standing; what is enough for a tradesman is less than what would be required for a gentleman, which, in turn, is less than for a noble or a king. But, according to Wilkins, there is a sufficiency corresponding to each social level, and what is enough at each rank is just so much and no more. This notion of limits is crucial because the observance of these limits will preserve the social order, whereas the opposite, the insatiable appetite for more and more wealth, will throw society into chaos.

For Wilkins, the principal agency for promoting sufficiency and discouraging reckless avarice is true religion. It is religion that teaches diligence, sobriety, and thrift, virtues that help human beings to achieve "a comfortable subsistence" suitable to the social status of each; religion then further teaches them to be happy with just that much and no more. Wilkins seems to have lived by his creed. At his funeral, William Lloyd (1627–1717), later a latitudinarian bishop, said of him:

> He knew the use of an estate, but did not covet it. What he yearly received of the Church, he bestowed in its service. As for his temporal estate, being secured against want, he sought no farther; I have heard him say often, *I will be no richer*, and I think he was as good as his word.[1]

Not only does religion produce good (moderate) people; it also produces a prosperous society: "Such places, where men have the opportunity of being instructed in, and excited to the duties of religion, do thereupon thrive and flourish most." Religion makes us both prosperous and temperate with regard to our getting and spending. In contrast to the extremes of wealth and poverty, which marred ancient times, "now there is employment enough for all, and money little enough for everyone." Religion has promoted the goal of economic sufficiency in which large numbers of people can find a comfortable subsistence. So "there is now a greater equality amongst mankind," that is, "equality" tied to what Wilkins meant by the pursuit of "absolute riches," enough to meet the occasions associated with one's social position and no more (quoted in Jacob 1992, 518–21).

Religion also promotes something else equally important, which we have, so far, noted only in passing: It is "one property of religion to civilize men, and make them more inquisitive in learning, and more diligent in practicing their several professions." With "a greater equality," religion also promotes practical, scientific knowledge. So "now the flourishing of arts and sciences hath so stirred up the sparks of men's natural nobility, and made them of such active and industrious spirits, as to free themselves in a great measure from that slavery, which those former and wilder nations were subjected unto." According to Wilkins, the world, with England leading the way, is on the threshold of the achievement of an ideal harmony—history is moving in the right direction—in which "the flourishing of arts and sciences" will spur human beings to greater and greater private industry, whose outcome will be private wealth distributed to provide a sufficiency proportioned to social station. And the engine driving the whole process is true (that is, Wilkins's) religion (Jacob 1992, 518–21).

There will be other benefits as well. Science will lead to the invention of labor-saving devices, which, in turn, will reduce "the hardness of toilsome labor and save much time,"[2] and the increased leisure can then be devoted to scientific and religious studies. (This was a pious and decidedly clerical view of how workers should spend their spare time.) One of Wilkins's colleagues at Oxford and in the Royal Society, William Petty (1623–87), worked out an economic theory for sustaining such a social system and agreed in 1664 that the leisure generated by increasing wealth and productivity should be spent on doing more science:

> What then should we busy ourselves about? I answer, in ratiocinations upon the works and will of God, to be supported not only by the indolency, but also by the pleasure of the body; and not only by the tranquillity, but serenity of the mind; and this exercise is the natural end of man in this world, and that which best disposeth him for his spiritual happiness in the other which is to come. The motions of the mind, being the quickest of all others, afford most variety, wherein is the very form and being of pleasure. (quoted in Hull 1899, 1:119–20)

This is a passage that repays close reading. The reference to "indolency" and "tranquillity" is a reference to Aristotelian values. Petty adds the pursuit of Epicurean "pleasure" to "indolency" and the pursuit of Stoic and Epicurean "serenity" to "tranquillity." In all this, he echoes the contemporary revival of interest in, and revaluation of, Epicureanism and Stoicism, and a corresponding depreciation of Scholastic teachings, in some circles in France and England. Petty had spent some time working as Hobbes's assistant in Paris, where Hobbes knew Mersenne, Gassendi, and Descartes, whose Christian Epicureanism and Stoicism we have already discussed.

Their Answer to Hobbes and the Radical Sects

Wilkins and his colleagues were deeply pious, social idealists. They were committed to a kind of science that would help them achieve their social and religious aims, that is, to building a particular kind of scientific and civil polity (Tuveson 1949). As such, their thinking represents a continuation of the utopianism that Bacon had initiated and that flourished in Hartlib's circle—but with this one big difference: Those earlier thinkers, like Hobbes, saw a powerful state, with virtually total authority over even the property and religious opinions of its subjects, as being the principal instrument for fostering the changes that would lead to the realization of their goals. It was this absolutism that Wilkins and the others now disavowed and, in its place, argued for self-discipline and self-help, the cultivation of individual virtue and piety, grounded on an optimistic estimate of man's intellectual and moral capacities that rejected the views of Hobbists and radical sectaries alike.

Human beings should not have to be subjected to Hobbes's state, according to Wilkins; most people are too capable of reason and virtue to need such a straitjacket. But the sectaries err on the other extreme when they claim that the saints, men and women inspired by God, are so capable of perfection that they should be left free to chart their own course. Despite their capacity for good, human beings are still not virtuous enough, according to Wilkins, to live without the restraints afforded by a society based on status and hierarchy. Instead, they must learn the discipline and self-restraint prescribed by the principle of sufficiency. Wilkins, in his own life, set the example to be followed. Finally, it is a combination of religion and science that teaches people how to observe and enjoy those natural limits so fundamental to personal happiness and public order.

The political thinking of Wilkins's protégé, Matthew Wren (1629–72), a fellow of the Royal Society and the cousin of the distinguished architect and astronomer Christopher Wren (1632–1723), provides a key to understanding this distinctive social theory. Wren's views are based on contract theory but are not Hobbist. In fact, they recall Gassendi's political thought. Individuals, Wren argued, should enter into a contract to set up a strong government, which is necessary to protect the obedient majority against the criminal element, a perennial minority who are prone to breaking the law and fomenting disorder. But the government must not be too strong, that is, so oppressive as to lord over it and coerce the peace-loving majority. The majority want a government effective enough to maintain the law and order that will provide the conditions in which they can devote themselves to earning a livelihood. But they do not want a government that uses its power to inhibit them in this enterprise. In fact, the role of the state must be to foster the industrious labor and prosperity of its private citizens and then to leave them to get on with their work.

What would happen if the majority's avid pursuit of gain went so far as to lead to a breakdown of the law and order on which that pursuit is predicated? In other words, what would happen if this desire for gain got out of hand, as Hobbes said it would in the absence of Leviathan, and destroyed the civil peace? For Wren, who developed his argument at Wilkins's request, such a possibility would be remote because most men are driven not by insatiable (Hobbist) greed but by "a desire of moderate riches, such as are subservient to the necessaries and conveniences of man's life, or to the attaining innocent and honest pleasures." Wilkins agreed with Wren that the majority, "the generality of a people," are rational enough to learn, with the help of the church, to exercise such self-restraint, that Leviathan is therefore uncalled for, and that social and political order can indeed be founded on the principle of sufficiency and on the basis of limited state power (Jacob 1992, 514–8).

For the latitudinarian natural philosophers, it was possible to have one's cake and eat it too, that is, to grow virtuous and prosperous at the same time; in fact, to do well by doing good. This would happen, they argued, not through massive government control but through strenuous self-exertion, not through utopian statism but through utopian voluntarism. Of course, "doing well" did not mean, to the latitudinarians, what it means today. Instead, it meant setting limits to getting and spending because the alternative, letting it rip, was fraught with danger—instability, disorder, faction, sedition, and civil war. Their thinking, then, was not capitalistic but antithetical to it and a response to all the "excesses" in word and deed, some of it "capitalistic" and some Hobbist and Plattist, that they were repelled by in revolutionary and postrevolutionary England. Perhaps, to some degree, they were salving their consciences, rationalizing their own acquisitive behavior, and refusing to face up to what Hobbes was telling them about themselves, though it would be difficult to make this case stick against someone like Wilkins. It was not until the 1680s that this Wilkinsian vision began to fade, but it took a new lease on life after the Revolution of 1688–9 (Jacob 1992; Jacob 1991).

Boyle

Robert Boyle (1627–91), perhaps the most important natural philosopher in the Wilkins group at Oxford and in the early Royal Society, clearly reflects both this new social idealism and the shift away from absolutism. The youngest son of the Earl of Cork, Boyle, in the 1640s and 1650s, adopted a social outlook similar to that of Wilkins and Petty. Influenced by Hartlib, he complained, like Plattes, of idle landlords and gentlemen (Boyle himself was a gentleman workaholic) and attacked covetous, ambitious men whom he called "Macchiavillians." Idleness and the desire for excessive wealth were impious, and offenders could be won back from such sins, and others strengthened against committing them, by a combination of science and religion, the worship of God, that is, through the study of his work.

Natural Religion

Boyle's argument runs as follows. From the Hermetic tradition and other sources, Boyle borrowed the image that served as the starting point of his natural religion: "If nature be a temple, man sure must be the priest." All creatures embody evidence of God's glory, but man alone is endowed with the gift of reason and so, being "the most obliged and most capable" creature, "is bound to return thanks and praises to his Maker, not only for himself, but for the whole creation." Not only is man "obliged" and "bound," he is also preprogrammed by God to carry out this mission: "To engage us to an industrious indagation [investigation] of the creatures . . . God made man so indigent and furnished him with such a multiplicity of desires; so that whereas other creatures are content with those few obvious and easily attainable necessaries that nature has almost everywhere provided for them; in man alone, every sense has store of greedy appetites" (quoted in Jacob 1992, 522).

But this universal human drive does not lead Boyle down the path to Hobbes's Leviathan. Our attempt to satisfy our desires, instead of making for anarchy and calling for the extreme remedy of Hobbes's sovereign, leads us to the discovery of nature, which not only provides knowledge but teaches virtue. We are not born wise and good, but we are wired by divine plan to become so. All that is needed is the requisite industry, which itself pays off by curing idleness and "sensuality." The result of all this virtuous industry is more virtue because nature revealed is a great moral teacher, presenting to the student a model of harmony with God's purpose: Each creature functions in such a way as to conduce to the end of all. The lesson is clear. This is the model that we would do well to follow. As Boyle says, "We might learn thence as much a solider as innocenter prudence then from the Books of Machiavell" (quoted in Jacob 1992, 523). The state is not the instrument of Boyle's moral reformation. Like Wilkins and Wren, Boyle makes us responsible for reforming ourselves, with considerable help, of course, from God.

Boyle's agreement with Wilkins and Wren is close, and because Boyle joined them at Oxford in 1655 or 1656, where he resided until 1668, there would have been ample opportunity for mutual influence. In a work first published in 1665, Boyle said, for instance, "A great or rich man's mind" can be "distempered with ambition, avarice, or any immoderate affection." Therefore: "Let us . . . consider both that fortune can give *much*, and [that] it must be the mind that makes that much *enough*." When men become "too greedy of superiority in fame and power," desire "degenerates into ambition." And the results are seditious:

> How many vices are usually set at work by this one passion! The contempt of the laws . . . and all other crimes and miseries that are wont to beget and attend civil wars, are the . . . dismal productions of this aspiring humour in a subject. (quoted in Jacob 1992, 523–4)

But preventive measures can be taken: "The usefulness of the passions should not hinder us from . . . employing the methods . . . afforded us by reason and religion to keep them within their due bounds." Among these "methods" is what Boyle calls "a competency of estate" proportioned to a man's "needs and conditions," which are, in turn, determined by "a man's particular quality or station." Boyle's agreement with Wilkins and Wren in these matters is complete, and, like Wilkins, he provides a religious foundation for the ethics they all three share (Jacob 1978; 1992).

Like Wilkins, too, Boyle was concerned, up through the early Restoration, with the threat to social order and private property presented by the radical sectaries, some of whom Wilkins referred to, in 1654, as that "gang of the vulgar Levellers" (Debus 1970, 200). In 1660, Boyle's close friend Peter Pett (1630–99) published a tract conceived jointly with Boyle and carefully vetted by him before publication and thus reflective of his views at the time (Jacob 1978, 133–4). Pett argued that radical millenaries "who would disturb civil Societies . . . by destroying propriety [property]" should be dealt with "as enemies of mankind." Boyle and Pett agreed with their Oxford colleagues, Wren and Wilkins, that steps should be taken to protect private property while maximizing trade and industry (Jacob 1978, 133–4; Pett 1661, 10–2).

Corpuscular Philosophy

Not only was Boyle an influential moral philosopher who shaped latitudinarian natural religion, he was also a great natural philosopher who shaped English scientific theory and practice. He had read Galileo's work in Florence, while on the grand tour in 1642. Returning to England, he was introduced to Hartlib in London and shared a strong commitment to chemistry and Baconianism with others in his circle. In the mid-1650s, he moved to Oxford at Wilkins's behest and pursued his research there for many years before returning to London, where he remained active in the Royal Society and in latitudinarian circles to the end of his life.

Early in his life, he read Gassendi, Descartes, and the Paracelsians and devoted himself to conducting chemical experiments and devising his own atomic or particulate theory, which he was careful to refer to as the corpuscular (rather than mechanical) philosophy. He was concerned to use this framework to invalidate Aristotelian and Paracelsian explanations for chemical phenomena. In his characteristically roundabout prose, he said,

> I hoped I might at least do no unseasonable piece of service to the corpuscular philosophers, by illustrating some of their notions with sensible experiments, and manifesting, that the things by me treated of may be at least plausibly explicated without having recourse to inexplicable forms, real qualities, the four peripatetick [Aristotelian] elements, or so much as the three [Paracelsian] chemical principles. (quoted in Hall 1981, 224)

Boyle's insistence on calling his physical theory corpuscular and not mechanical was more than a matter of semantics. He used his corpuscular philosophy so that phenomena might be "at least plausibly explicated." He aimed not at reaching the final truth but at obtaining certainty, that is, a plausible, provisional explication. In this respect, he was much closer to Mersenne and Gassendi than to Descartes. In fact, Boyle was as unsympathetic to Cartesian mechanism as to any other kind of dogmatism. He felt that any attempt at system-building was reductive, premature, and counterproductive, contrary to the humility required for effective experimental inquiry. In this respect, he harked back to Bacon: Nature must be obeyed before it can be commanded (Kaplan 1993, 44–61, 68–76; Sargent 1995).

Boyle referred to matter and motion as "the two grand and most catholick principles of bodies." But, in his view, many chemical processes could not be reduced to simple mechanical terms, the result of particles bouncing off, or rubbing against, one another, as a Cartesian would claim. Instead, he attributed to some corpuscles specialized functions that were not the result of purely mechanical motion. Thus, he did not hesitate to refer to the role played by Paracelsian volatile spirits, active substances, or seminal principles in certain kinds of chemical change. Moreover, "a chemical reaction was for Boyle a realignment of the component particles or corpuscles . . . of the reagents, which he came near to attributing to variations of affinity between them"—again a phenomenon not reducible to mechanical terms. The corpuscular philosophy was an excellent hypothesis, perhaps the best, but it was still only a working hypothesis and, as such, did not provide a universal explanation of nature but a useful framework for exploring it (Westfall 1977, 77; Hall 1981, 230, 232).

From reading Gassendi and others, Boyle was favorably disposed to ancient Greek atomism. But, on one point, he was adamant in his opposition to it: Motion must not be thought of as inherent in matter either from all eternity, as Epicurus and Lucretius had claimed, or as imparted by God at the moment of creation, as Walter Charleton (1619–1707), the translator of Gassendi's works into English, was then arguing. The world, according to Boyle, does not run on its own even after God has set it in motion at the moment of the creation; it remains ever susceptible to God's absolute power to intervene. In fact, Boyle and fellow latitudinarians like Wilkins and Joseph Glanvill were so anxious to defend the role of God's sustaining Providence in the universe, against the threat posed by Hobbists and extreme Cartesian mechanists, that they investigated (and took seriously) reports of instances of witchcraft. For Boyle, Glanvill, and others, such phenomena constituted strong empirical evidence for the operation of supernatural spirits, good and bad, in the created world and thus furnished an argument against a purely mechanistic account of nature (Westfall 1977, 75–81; Clericuzio 1990; Webster 1982, 88–103).

On the other hand, when it suited Boyle's purposes, he could incline to a strictly mechanical theory of chemical change. Thus, deploying the mechanical hypothesis, he argued that almost anything can be made from anything, an assertion that served to underwrite his belief in alchemy. For many years, he searched for ways to transmute gold and traded alchemical secrets with other thinkers like John Locke (1632–1704) and Isaac Newton, who were also avid alchemists. There were times, too, when Boyle described his philosophy in explicitly mechanical terms:

> The universe being once framed by God, and the laws of motion being settled and all upheld by his incessant concourse and general providence, the phenomena of the world thus constituted are physi-

> cally produced by the mechanical affections of the parts of matter, and what they operate upon one another according to mechanical laws. (quoted in Shanahan 1988, 567)

But even here at his most "mechanical," the reference to God's "incessant concourse" gives Boyle away: Motion is not inherent in matter but continually imparted by God. Boyle's version of mechanical philosophy is not Cartesian but something much more provisional, which might best be called experimental, corpuscular philosophy (Kuhn 1952; Henry 1986; Shanahan 1988).

Experimentalism

Boyle was an experimentalist who believed, like Petty and others in the Wilkins circle, that the way to scientific knowledge lies not in academic debate, such as Scholastics engaged in, nor in secret revelations from on high, such as extreme Paracelsians claimed, nor in the building of logically deductive, metaphysical systems, whether Aristotelian, Cartesian, or Hobbist. Rather, the surest road to discovery, for Boyle, lies in experimentation. As his colleague Petty is supposed to have said to Hartlib, "Whoever can make out by visible experiment that which he undertakes, he will evince that he hath the truth. For all knowledge must be brought before their true judges which are either demonstration or sense and experiment, the rest is mere wit or rhetorick."[3] Boyle devoted himself to careful, exhaustive inquiry. He consulted the experts—not just academics and scholar-gentry but artisans, travelers, and men of affairs, whose information derived from experience. He also spent much of his life conducting experiments and made his reputation for some famous experiments using the air pump (Sargent 1995). Before discussing these, we shall first sketch in the important experimental background to Boyle's efforts.

The conception of nature that owed so much to Galileo and Mersenne held that the material world can best be understood by measuring it, that the most valid explanations of natural phenomena are the ones that rely on quantification to account for observed behavior. The effectiveness of this approach was clearly demonstrated in a number of experiments performed with the aid of a new device known as the barometer. A prototype that used water was built in Rome in the early seventeenth century. A glass tube, with one end open and the other end closed, was filled with water and suspended vertically in a pan of water, with the open end submerged in the water. This simple device demonstrated two things, each of which prompted a question. First, a space was left between the top of the column of water and the closed end of the tube: What, if anything, was in there? Second, water could be suspended in the tube to a height of almost thirty-four feet: why no more and no less?

Answering the second question, some observers claimed that the water rose almost thirty-four feet because, at that height, the weight of the water in the tube equaled the weight of the atmosphere on the outside, which sustained it. This strictly physical explanation was confirmed in 1644 in an experiment made by Galileo's Italian follower Evangelista Torricelli (1608–47). He built a barometer that substituted mercury for water; in fact, it was the first mercury barometer. Mercury is fourteen times heavier than water. If the ascent of a fluid in a tube is the result of a quantifiable, physical factor, that is, outside air pressure, mercury should rise only one-fourteenth as high as water—and so Torricelli found that it does.

As for that first question—what is in that space between the fluid and the top of the tube?—opinion was deeply divided. Some philosophers were inclined to argue that the space is empty and constitutes a true vacuum. But Aristotle and his followers held that "nature abhors a vacuum," that matter always rushes in to fill empty space, not mechanically but animistically, as if brute matter is guided by instinct to preserve the universal plenum. Descartes was also a plenist, but of a different sort. The universe, he held, is made up of two substances, corporeal and incorporeal, body and soul. By definition, bodies are extended, souls unextended. Empty space, incorporeal extension, is therefore a contradiction in terms, and a vacuum cannot exist in nature. What, then, is the status of the space at the top of the barometric tube? Plenists, whether Aristotelians or Cartesians, maintained that this space only looks empty but must really be full. Perhaps, they argued, it is full of vapors given off by the liquid, and the vapors drive it down when it would otherwise rise to the top of the tube.

But Blaise Pascal (1623–62), the French mathematician and philosopher, exploded that theory by conducting a public experiment using two barometers, one of water and the other of wine. Wine was known to be more vaporous than water, which suggested that, if the plenists were right, the wine should be driven down farther than the water. On the other hand, wine is lighter than water, which suggested that, if the vacuists were correct, the wine should rise higher than the water. Once performed, the experiment corroborated the existence of a vacuum, and the plenists were defeated in full view of an audience who had gathered to witness the outcome.

Even more famous in the history of science is the barometric experiment Pascal asked his brother-in-law to perform in 1648 on the mountain called the Puy de Dôme in east central France. As the barometer was carried to the summit, the height of the column of fluid fell because the weight of the air above it dropped, and the conclusion was thus confirmed that what the device registers is a simple balance of weights, the weight of the atmosphere pressing down from the outside and the weight of the column of liquid inside the tube. Dogmatic explanations of why vacuums cannot exist, whether Aristotelian or Cartesian, were refuted. Victory belonged to the view that natural phenomena can be understood to the extent that they can be quantified and those quantifications confirmed by experimental investigation (Westfall 1977, 43–8). Pascal put it this way:

> Let all the disciples of Aristotle . . . learn that experiment is the true master that one must follow in Physics; that the experiment made on the mountains has overthrown the universal belief in nature's abhorrence of a vacuum; . . . and that the weight of the mass of the air is the true cause of all the effects hitherto ascribed to that imaginary cause. (quoted in Hall 1981, 253)

Of course, the experiment told against the dogmatic Cartesians as well.

Boyle was heir to this legacy. His anti-Aristotelian reasoning is particularly interesting. Not only did he accept the mechanical explanation for liquid rising in a tube and reject the animistic one, but his reasons for doing so were partly theological. To claim, as the Aristotelians do, that nature abhors a vacuum, Boyle said, is to endow brute matter with foresight and purpose; it is to assert

> that a brute and inanimate creature, as water, not only has a power to move its heavy body upwards, contrary (to speak in their language) to the tendency of its particular nature, but knows both that unless it succeed the attracted air, there will follow a vacuum; and that this water is withal so generous, as by ascending, to act contrary to its particular inclination for the general good of the universe, like a noble patriot, that sacrifices his private interests to the publick ones of his country. (quoted in Jacob 1978, 113)

For Boyle, this kind of thinking was tantamount to putting the whole of nature on the same level with human beings endowed by God with the gift of reason. This is the gift, according to Boyle and his latitudinarian colleagues, that sets us above other creatures by giving us the capacity to make moral choices, to choose right over wrong, and, thus, to win eternal life. But to put ordinary matter on a rational par with us is to raise the question whether we are not like the so-called lesser creatures and therefore not immortal, to insinuate that we may live and die like the rest of creation, without hope of salvation. Here was a threat that struck at the roots of Boyle's latitudinarian moral theology, at the doctrine of rewards and punishments meted out in the afterlife and at the Christian moral order based upon it.

Boyle saw that strictly mechanical explanations of natural processes overcame this threat by preserving the proper distinction between rational humanity and stupid matter. Descartes, as we have seen, deployed his dualistic metaphysics, sharply dividing reality between body and soul, to answer the same threat (pp. 80–81). The threat itself echoed the heretical naturalism and pagan vitalism associated with Pomponazzi, Bruno, and Vanini that Mersenne was so concerned to refute. It is significant that Boyle and Descartes also framed their respective natural philosophies in this context and in answer to this same heretical challenge. In Boyle's case, the threat may also have hit closer to home. Some radical sectaries preached that the soul was mortal and died with the body, and Winstanley, the Digger leader, conflated God and nature (Jacob 1978, 113–5; Hill 1972).

By 1654, the air pump had been invented in Germany by Otto von Guericke (1602–86). Boyle immediately saw the experimental possibilities opened up by the new device and had one constructed for his own use by his gifted assistant Robert Hooke (1635–1703), who later became curator of experiments at the Royal Society. Using the pump, Boyle conducted a brilliant series of experiments on the physical properties of air. Going beyond earlier findings as to the weight of the air, he demonstrated, for instance, that light, but not sound, can be transmitted through a vacuum, that a flame is extinguished and animals and plants perish without air, and that air is elastic. This last, the elasticity of the air, means that air particles behave like tiny coiled springs that exert pressure, and do so in a regularly variable way relative to the amount of space they occupy. This last finding led him to the "law" that bears his name, Boyle's law, which states that air exerts a pressure that varies inversely with the volume of space it occupies. Like Pascal, Boyle established a significant physical relationship on the basis of tabulated experimental data; his achievement represents another triumph for the empirical method, quantification, and a mechanical conception of nature. Boyle, true to his diffidence about drawing too general a conclusion from his researches, never used the word "law" to describe this particular finding (Hall 1981, 255; Shapin and Schaffer 1985).

But, if his claims were always modest, his experimental program was not. Boyle chose big subjects for investigation, like the air, which, he said, was "not, as many imagine, a simple and elementary body, but a confused aggregate of effluviums from such differing bodies, that... perhaps there is scarce a more heterogeneous body in the world" (quoted in Kaplan 1993, 89). Of course, the air pump gave him the instrument for conducting experiments, and the high level of interest, at the time, in the phenomena of heat and respiration stimulated his inquiries. William Harvey had recently discovered the circulation of the blood, and Harvey's successors (Boyle's Oxford colleagues) were concerned to explore the role of air in animal physiology. Boyle participated in this research. He tried to explain breathing in terms of differences in air pressure outside and inside the lungs. When he found that both life and flame are extinguished in a vacuum, he speculated that air is necessary to the process by which it was believed the heart generates the heat essential to life. He also thought that chemical analysis of streams of particles carried in the air might lead to knowledge of the causes of epidemic diseases and to new drugs for treating them (Kaplan 1993).

He was not a mere collector of information but worked out careful procedures for conducting experiments that were designed to test hypotheses. In his reports, he never claimed more than he could demonstrate, and he described experiments that failed. This last habit was essential to progress in experimental philosophy, he argued, because it exposed the limits of a hypothesis and thus forced the investigator either to refine or abandon it. Boyle became the acknowledged master of one kind of scientific method identified with the early Royal Society. He was scrupulously empirical, reluctant to proceed from the data to make claims as to their causes, and unwilling to attribute more than probable certainty to his conclusions. It is often argued that, in matters of method, Newton was Boyle's follower, and, to some extent, this is true. But Newton aimed for a much higher degree of certainty than Boyle did; Newton's search for mathematical laws was an ambition Boyle eschewed on the grounds that its immodesty jeopardized the search for truth (Hall 1981, 320–1; Mandelbaum 1964, 88–117; Sargent 1995; Shapin and Schaffer 1985).

The Organization of Science

The Royal Society was at the center of scientific life in England. It developed out of the efforts of Wilkins and many others over the previous twenty years, but, dissociating itself from its Puritan and Cromwellian roots, it was self-consciously royalist and received a charter from the recently restored king, Charles II, in 1662.

It met regularly in London and grew to well over two hundred members, who included both serious scientists, like Boyle, Hooke, and Newton, and untrained amateurs who enjoyed the meetings and basked in the prestige of the society. It was intended to be a national organization, but it excluded women, servants, and the poor and elected very few merchants to membership. Nor was it inclusive of every major scientific thinker. Hobbes, for example, was excluded for obvious reasons. Most members were both well heeled and well educated, men with enough time and money to buy, and read, scientific books, and to frequent the new coffeehouses, where science was discussed. The members, called fellows, associated the Royal Society with Salomon's House in Bacon's *New Atlantis,* and the rhetoric of apologists for the society was utopian and millenarian: Science would bring peace, prosperity, and Protestant victory in England and throughout the world, and, through science, the latitudinarian social and religious vision would be realized. The chief public advocates for the society—Thomas Sprat (1635–1713), Joseph Glanvill (1636–80), Samuel Parker (1640–88), John Beale (1608–83), and John Evelyn (1620–1706)—belonged to Wilkins's party in the post-Restoration English Church and promoted his views (Aarslef 1976; Jacob 1991; Shapiro 1968).

In one crucial respect, however, the Royal Society departed from the Baconian blueprint. For Bacon, Salomon's House was the most important government agency, a key part of the royal bureaucracy; but this was never to be true of the Royal Society. Though chartered by the king, it was not funded by the state at all during its first twenty years and only began receiving some regular income from a modest endowment the king finally bestowed in 1682. The fellows were naturally envious of the resources and research facilities provided by the French king to the newly founded Royal Academy of Sciences because the Royal Society, for its part, was always short of money.

Though hobbled financially, the society was nonetheless successful in promoting science and gaining prestige. At least there were sufficient funds to pay a secretary, Henry Oldenburg (1618?–77), and a curator of experiments, most notably Robert Hooke. The society's meetings, moreover, were conceived as the place where experiments should be performed in front of the members present. The idea was the legacy of Bacon, Mersenne, Wilkins, and Boyle. Knowledge would come not from metaphysical speculation or contentious argumentation but from a public witnessing of an experiment by trustworthy men—that is, gentlemen—from which consensus would emerge concerning the result. An experiment might succeed or fail in establishing anything conclusive. Either way, a consensus would be reached by an on-the-scene tribunal of socially elite observers. Such experiments were regularly conducted, especially during the first two decades of the society's existence. Later on, however, meetings were given over to reporting and discussing scientific information rather than to actual experimentation, which was conducted elsewhere. Boyle's London house, for instance, was fitted out with a well-equipped laboratory. He was one of those wealthy gents upon whom, in lieu of state funding, English science and the early Royal Society depended (Hunter 1981, 32–49, 65–86; Shapin and Schaffer 1985; Shapin 1988).

Hooke was responsible for demonstrating experiments at the weekly meetings of the society during its early years. He had a long and many-sided career. Not only did he have the practical skills needed to conduct experiments, he was an expert instrument-maker and invented improved versions of the telescope, barometer, and microscope. With this last instrument, he undertook the systematic investigations that led to his most famous work, *Micrographia,* published in 1665 under the auspices of the Royal Society and intended to enhance its reputation. Here, he revealed, in both words and graphic printed images, the exquisitely structured world too small to be seen by the naked eye (Fig. 14). Hooke went on to make contributions in a number of fields—clock making, the theory of light, and the theory of gravity—and, in this last field, his thinking may have influenced Newton. Hooke was also an accomplished architect. His career exemplifies the alliance between head and hand, abstract thinking and practical skill, theoretical science and applied science, which had been the features of intellectual activity at Gresham College that Wilkins so much admired. Wilkins supervised the production of Hooke's *Micrographia* (Hunter and Schaffer 1989).

The man most responsible for the success of the society as a center and symbol of scientific activity was Oldenburg. Though meagerly paid, he was tireless in his devotion to the cause. Like Mersenne and Hartlib before him, he carried on an enormous correspondence with practitioners

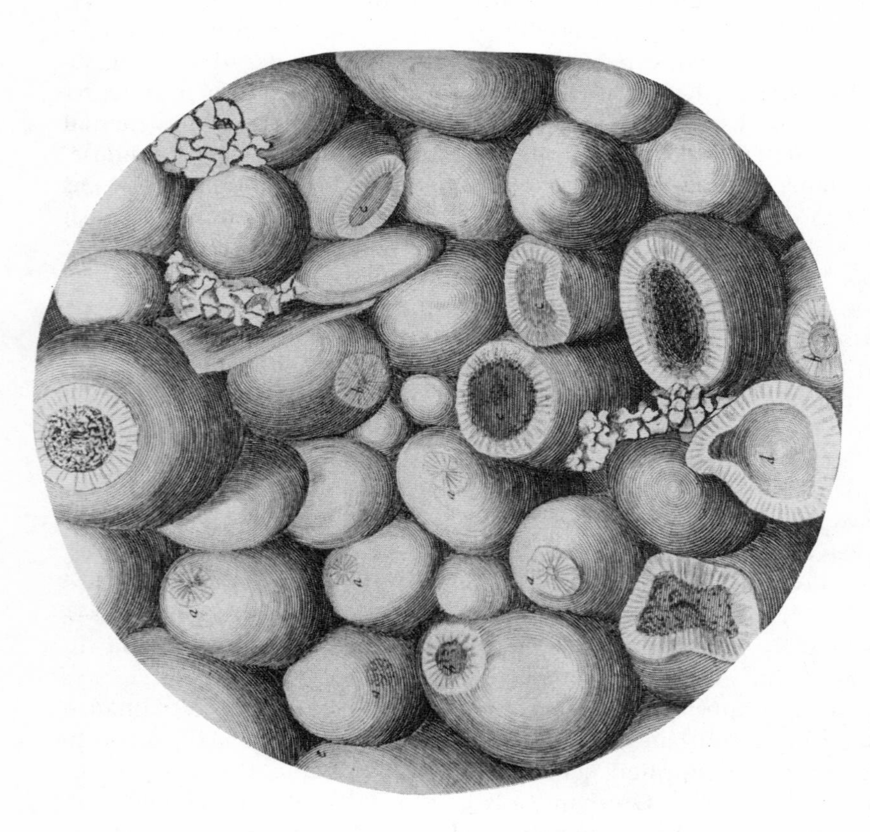

FIGURE 14 An Illustration of the Porous Surface of a Stone as Revealed Under a Microscope. (From Robert Hooke's *Micrographia*, 1665. Courtesy of Van Pelt Library, University of Pennsylvania.)

both at home and abroad, through which scientific knowledge was spread, further investigation stimulated, and the activities and mission of the society publicized. Equally important, he launched a new journal, the *Philosophical Transactions*, in 1665, and served as its editor until his death. Though initially his private project designed to supplement his income, the journal was, from the start, associated by its readers with the society itself. It published research reports and book reviews and was even more important than Oldenburg's correspondence was to the spread of scientific information and discoveries. It was through Oldenburg's efforts, and under the aegis of the society, that a national and international scientific community was knitted together for the first time (Hunter 1981, 49–58).

But the road was not smooth. Besides persistent money problems, the Royal Society had its vocal critics—including satirical wits who poked fun at it, pious Christians who believed scientific investigation detracted from the worship of God and a preoccupation with the natural world might lead to neglect of the supernatural, and, coming from the opposite direction, radicals like Hobbes and his anticlerical followers for whom organized science was a new kind of priestcraft destructive of civil peace. Against these various threats, the society stood its ground until, by the eighteenth century, faith in science as a servant of religion and progress, wealth and virtue, was becoming more and more widespread. What helped to turn the tide in this direction, perhaps more than anything else, was the enormous intellectual achievement of one man, Isaac Newton (Hunter 1981, 136–87; Shapin and Schaffer 1985, 80–154).

Newton

Early Development

Newton was closely associated with Cambridge University, where he was first a student, then a fellow of Trinity College and, from 1669, the Lucasian professor of mathematics, the post vacated by his former tutor, Isaac Barrow (1630–77). Newton was a solitary figure who devoted himself to his studies in a number of fields—including alchemy, biblical chronology, and church history, as well as the mathematics, optics, and astronomical physics in which he made his great contributions. As a student, he quickly absorbed the teachings of the leading contemporary natural philosophers in France and England, including Descartes, Gassendi, Boyle, Hobbes, Charleton, and Henry More (1614–87).

The basic framework of his early scientific thinking was provided by the mechanical and corpuscular philosophies, which exposed the deficiencies of Aristotelianism and taught that the better way to understand nature is in terms of moving particles of matter (Dobbs and Jacob 1995, 3–15).

But Newton was a devout Christian who seems to have always been alert to what he felt were the religious shortcomings of mechanism, to the fact that Cartesian metaphysics, for example, unduly restricted the power of God to act in nature and thus offered, as he put it, "a path to Atheism."[4] There was, throughout his career, a deep religious component in his scientific thinking. God was never to be left out of the picture and His role in the universe was always to be vindicated. Thus, like Boyle, Newton early believed that matter is inert and incapable of self-motion, that God is the ultimate source of all motion and order in the world. Later, Newton went further and drew from his lengthy alchemical studies and his reading of the ancient Stoics the view that there is, in nature, a "vital alchemical agent that acts in the formation of everything" and that, through this agent, God constantly molds the universe to His design (Dobbs and Jacob 1995, 27, 31).

Newton also believed, as early as the late 1660s, that infinite three-dimensional space is not only the scene of God's acting, the theater of His Providence, but that such space is itself uncreated and divine, that is, "God's property and, in effect, his immensity." In 1713, Newton said that God "endures forever and is everywhere present; and by existing always and everywhere, he constitutes duration and space." On this important point, Newton was following the Cambridge Neoplatonist Henry More, who, in the late 1640s, had welcomed Descartes's metaphysics but who, by the 1660s, had rejected the Cartesian material plenum because it left no room for the operation of spirit in the universe. Instead, More adopted the Stoic world-picture of our finite world surrounded by infinite void space. From there he went on, like Gassendi, to divinize this space, to identify it with infinitely extended spiritual substance, or God. In these matters, Newton was More's disciple. But, until the early 1680s, Newton was still enough of a mechanical philosopher to think that an ether of fine, moving particles is the cause of gravity and holds the planets in their orbits in a fashion reminiscent of the Cartesian vortices (Grant 1981, 225–61).

This problem of the planetary orbits remained the big question in physical astronomy. Copernicus, Brahe, Galileo, and Kepler had unhinged the planets and stars from their spheres and argued (but not proved) that the earth revolves around a stationary but rotating sun. How, then, can we explain why the heavy earth moves at all? (Aristotelians said that a moving earth is a contradiction in terms.) How further can we explain how the planets stay in orbit now that the perfect spheres, which used to be thought of as tracks, are gone? The Cartesian vortices offered one explanation, Newton's youthful ether another, and Kepler had suggested a combination of magnetism and an immaterial force emanating from the sun. But these were speculative answers that were not susceptible to mathematical measurement: They could be postulated but not demonstrated. So the question still begged to be answered, and it was Newton who finally did it.

The Newtonian Synthesis

Newton did not work in a vacuum, however. From the legacy left by Galileo, Kepler, Descartes, and Gassendi, he derived his solution. He was a creative synthesizer, in other words, and his synthesis was a feat of the greatest genius.

As noted in chapter 3, through observations, Kepler had found three laws:

1) the planets move in elliptical orbits around the sun,
2) in these orbits, they sweep out equal areas in equal times, and
3) the squares of their times around the sun are proportional to the cubes of their mean distances from the sun.

The mathematical regularities of Kepler's astronomy were inspiring. On the basis of his last two laws, Newton's associates in the Royal Society, Hooke, Christopher Wren, and the astronomer Edmond Halley (1656–1742) in particular, speculated that there must be a measurable force of gravity operating between the sun and each of the planets, which holds them in their elliptical orbits. They even supposed that such a force would be inversely proportional to the square of the distance between the sun and each planet. But, try as they might, they could not provide a mathematical demonstration of the operation of the force of gravity. So Halley turned to Newton for help in 1684, and, much to his surprise, Newton claimed to have already calculated gravity's force many years earlier. But, having mislaid the papers, he was to spend the next three years retracing his steps, working out, at Halley's request, the mathematical proof for the law of universal gravitation. In doing so, Newton provided a whole new framework for understanding the physical world, based on a mathematical explanation of the behavior of moving objects in space, whether on earth or in the heavens. The result of Newton's efforts was the publication, in 1687, of the *Philosophiae Naturalis Principia Mathematica* (*The Mathematical Principles of Natural Philosophy*), perhaps the most important scientific book ever written.

In it, Newton asserts the existence of a universal force of gravity operating between every body in the universe and every other body. The same force that famously causes apples to fall from trees helps to keep planets in orbit around the sun. The earth attracts the apples, the sun likewise the planets. Galileo had calculated the rate of acceleration at which things, like apples, fall to earth; Newton found that the same calculation applies to the planets in the heavens. The nature of gravity is everywhere the same and can be stated in terms of a mathematical law, whereby every body in the universe attracts every other with a force directly proportional to the product of the two masses and inversely proportional to the square of the distance between them. This law is universal, which means that there is only one physics for earth and heaven and that the old binary world-picture based on Aristotelian tradition was finally disproved and replaced.

But, of course, the movement of the planets is more complicated than that of apples dropping from a tree, because planets do not fall but instead describe Kepler's elliptical orbits. They do so because they have two motions, a continual falling motion toward the sun combined with an inertial motion in a straight line tangential to their orbits around the sun. From Descartes, Newton borrowed this idea of inertia, which became, for him, a fundamental law of motion: Every body will continue in its state of rest or uniform motion in a straight line unless compelled to change by external forces impressed upon it.

Before such inertial motion could make sense, however, something else was needed, namely, a universe in which a finite quantity of matter could be seen as moving through infinite, absolute space. How else could inertia be seen to work? How could an object move in a straight line forever unless the space through which it moved were infinite and absolute? Descartes had denied the existence of such space, but Gassendi and More had asserted it, and so Newton borrowed just the world-picture he needed from them, who, in turn, had taken it from the ancient Stoics (Grant 1981, 241–4).

Material objects, including the planets, could now be conceived of as endowed with inertial force and as moving through infinite space in a straight line forever, unless, or until, acted on by other forces, one of which is gravity. The planets, thus conceived, would soon leave the solar system and sail out into limitless space were it not for the pull of gravity. In fact, their elliptical orbits are precisely the product of the balance between these two forces—(1) gravity, by which they perpetually tend to fall into the sun, and (2) inertia, by which they perpetually tend to fly out into space. It was Newton's incomparable genius to be able to show this in mathematical terms and so to synthesize the Cartesian concept of inertia with Galilean terrestrial physics and both of these with Kepler's three laws of planetary motion—all within the framework of a neo-Stoic cosmos updated by Gassendi, More, and, finally, Newton himself (Dobbs and Jacob 1995, 38–46; Cohen 1960, 152–90; Westfall 1980).

Newton's early and midcareer (1665–1704) was punctuated by notable achievements, among which three more will be mentioned: (1) in mathematics, the discovery of calculus in 1665–6, an honor he shares with his contemporary Gottfried Wilhelm Leibniz (1646–1716), the German philosopher who also independently discovered it; (2) the invention, in 1668, of the reflecting telescope, which uses mirrors rather than lenses to concentrate light and obtain magnification; and (3) another great book, the *Opticks* (1704), based on brilliant experimental work, in which Newton worked out a new corpuscular theory of light.

But What Is Gravity?

Newtonian physics is a triumph of mathematical demonstration. But there was still a big question. Gravity can be measured mathematically, but what is it substantially? The Cartesian mechanists, led by Leibniz, ridiculed Newton's gravity for reintroducing into natural philosophy this mysterious action at a distance. It smacked to them either of the occult qualities associated with a discredited Scholasticism or of a new kind of magic so recently banished by Descartes. But Newton and his latitudinarian followers were quick to take up the challenge and to insist that gravity, whatever it is in and of itself, demonstrates the power of God acting in the universe and sustaining the order of nature. For the devout, the law of universal attraction was tantamount to proof of divine Providence, a great victory for true Christianity and a blow struck against its enemies, especially radical freethinkers influenced by Hobbes and an unorthodox reading of Descartes (Koyré 1968, 155–89, 206–20, 235–72; Jacob 1991, 162–200; Descartes 1968, 64–5).

Many of these freethinkers were deists who argued that the world-machine can run on its own because God has created it perfect in the first place. Other freethinkers were pantheists who, harking back to Bruno and Winstanley, maintained that God and nature are one, and put their own radical spin on Newton's gravity. Their argument was as simple as it was subversive. Who can claim that gravity is a force imparted by God, when one may just as logically say that it is a property of matter itself? And so they did. Contrary to the view of Newton and the orthodox Newtonians, matter is not dead and lifeless, created and acted upon by divine agency, but alive and uncreated, eternal, infinite, and intelligent. The material universe, in other words, organizes and governs itself, and the force of gravity inherent in matter is what we understand of this self-governing capacity. Both deists and pantheists enlisted Newtonian gravity to support their case that nature can be explained in terms of matter in motion alone, without reference to divine Providence. If this is so, they went on to say, what need is there of the clergy or the Bible to explain the ways of God to humankind, especially when science can do the job without resorting to miracles and mystery (Koyré 1968, 273–6; Jacob 1991, 201–50)?

Newtonian science can be said to have produced a dual legacy, divided between those who used it to uphold orthodox Protestant Christianity and those who used it to tear it down. The battle lines were thus drawn and the battles fought during the eighteenth-century Enlightenment, fueled in part by Newton's achievement and one more measure of its enormous and lasting impact. With respect to this division, Newton's own position was ambiguous. On the one hand, he urged his latitudinarian followers to use his science to defend the church against the freethinkers, "to come," as he said, "to the knowledge of a Deity by the frame of nature." This, after all, was what the ancients had done before religion was corrupted. On the other hand, he was a freethinker of sorts himself because he was a secret (but not a public) anti-Trinitarian: God is one, not three-in-one, as orthodox Christians held. Newton demoted the Son in favor of the Father. But it seems that Newton's social conservatism won out over his theological scruples. A strong state church, even though doctrinally unsound, was preferable to organized religion besieged, and perhaps destroyed, by secularizing freethinkers (Dobbs and Jacob 1995, 46–71, 95–104; Koyré 1968, 273–6).

Curiously, Newton's anti-Trinitarian beliefs may have nudged him to rethink the problem of the cause of gravity. He was long persuaded to think that it was caused not by anything physical but by something he derived from his reading of the ancient Stoics, which he called an "active principle." In Newton's hands, the subtle matter, or breath, of the Stoics was spiritualized and became the divine, immaterial substance, permeating everything, aware of everything, and directly responsible for universal gravitation. But Newton's anti-Trinitarianism led him eventually to believe that the Supreme Deity was transcendent, completely removed from the day-to-day operations of the universe. So Newton sought another explanation for gravity that would save God's transcendence. As he watched Francis Hauksbee (1670–1713) conduct experiments on electricity at the Royal Society, he thought he might have found what he was looking for. Electricity, so elastic as to fill universal space but so weightless as to offer no resistance to the planets, might be the halfway house between body and spirit that causes gravity while preserving God's transcendence (Dobbs and Jacob 1995, 46–56).

The legacy of Newtonian science also had a practical side. John Theophilus Desaguliers (1683–1744), the early eighteenth-century Newtonian, was an Anglican parish priest but notoriously neglected his parishioners in favor of his devotion to experimental science based on the works of Boyle and Newton. He lectured widely on practical science and was concerned to establish what he regarded as a vital link between science and engineering. Echoing Boyle, he believed that scientists can profit from what skilled craftsmen have to teach them. Conversely, he insisted that artisans and engineers should learn science in order to base their practice on proper theory. Desaguliers also published his lectures in *A Course of Experimental Philosophy* (1734), which was to be used as a textbook in applied science for many decades. The book began with an exposition of the principles of mechanical science drawn from the works of Boyle and Newton. Later chapters were devoted to the practical application of this theoretical knowledge to improving machines, including the steam engine, and to the commercial uses to which they could be put. Desaguliers's outlook, and that of many others like him in England, has been called "proto-industrial," and, as such, was an important ingredient in setting the stage for the first Industrial Revolution (Dobbs and Jacob 1995, 71–95).

Conclusion

Science in seventeenth-century France and England, as we said in the introduction, developed in response to two major stimuli, the search for truth about nature, and the religious beliefs and commitments that helped to motivate that search and gave it added meaning. As we have seen in the last two chapters, things turned out differently between the two kingdoms. In France, the result was the triumph of Cartesianism and the creation of the Royal Academy of Sciences; in England, the outcome saw the triumph of Newtonianism and the emergence of the Royal Society. But, beneath these basic differences lay a common aspiration that has often been overlooked but is well worth pointing out.

Natural philosophers, both French and English, aimed not only at the discovery of truth about the natural world and the mastery of nature for material human benefit. They also looked forward to the progressive purification of religion and a moral reformation, both individual and collective, to which science, that is, the right science, was seen as being crucial. Through science, so it was thought, people were not only expected to know more and to live more healthily and securely, they were also expected to become more civilized, to live more virtuously and civilly, in better order and greater comity. What fed this search for moral order was some combination of hopes and fears—fears of heresy and religious conflict, moral decay and social breakdown, and, conversely, hopes born of scientific discoveries, biblical prophecy, and social idealism of one kind or another. In England, a millenarian thrust contributed to this mixture. Among natural philosophers both in England and on the Continent, this search for moral order led to a remarkably similar intellectual result. French and English thinkers—Gassendi, Descartes, Wilkins, Boyle, and others—committed themselves to moral theologies founded in part on a Christian humanist tradition and in part on ancient classical sources, especially Stoic and Epicurean teachings, now, of course, made compatible with Christianity, indeed made to further one or another version of Christian piety.

The moral theology that grew up in England around the central figures of Wilkins and Boyle was particularly distinctive and emerged during the midcentury upheaval in reaction to the Puritanism, radical sectarianism, and Hobbism that were thought to jeopardize true religion, social order, and private property. The result of this reaction would help to determine the meaning of the Scientific Revolution in England. Robert Merton argued in the 1930s that Puritanism and capitalism spurred the development of science in seventeenth-century England (1970, 55–136). But, as we now see, it was not Puritanism and capitalism that provided the ideological underpinnings of English science during the second half of the century. It was, rather, a reaction that set in among scientific thinkers to what they took to be the excesses of

Puritan enthusiasm and Hobbist materialism. Out of that reaction, Wilkins, Boyle, and their circle in the Royal Society carved a social outlook that defined and propelled their efforts to reform both science and morals, knowledge and behavior, in revolutionary and postrevolutionary England (Jacob 1992; Jacob 1991).

Notes

1. William Lloyd, *A Sermon Preached at the Funeral of... John [Wilkins]... Bishop of Chester* (London: By A. C. for Henry Brome, 1672), 29.
2. Sheffield University Library, Hartlib Papers, 50H 28/1/77b, "Ephemerides 1650."
3. Ibid., 28/1/10a, "Ephemerides 1649."
4. Quoted in J. R. Jacob and M. C. Jacob, "The Anglican Origins of Modern Science: The Metaphysical Foundations of the Whig Constitution," *Isis* 71 (1980): 262.

Selected Bibliography

WHAT FOLLOWS IS a list of basic works cited in the text. This is a highly selected list designed for an introduction to the subject, which is vast. Readers are referred to the bibliographies in the books on this list for further reading on any given topic and to C. C. Gillispie (ed.), *Dictionary of Scientific Biography*, 16 vols (New York: Scribner, 1970–80) for short, informative accounts of most of the thinkers treated in the text. There is an informative and up-to-date bibliographical essay on this large subject in Steven Shapin, *The Scientific Revolution* (Chicago: University of Chicago Press, 1996): 167–211.

Aarsleff, Hans. 1976. "Wilkins, John." In *Dictionary of Scientific Biography*, vol. 14, ed. Charles Coulston Gillispie. New York: Charles Scribner's Sons.

Allen, Don Cameron. 1944. "The Rehabilitation of Epicurus and His Theory of Pleasure in the Early Renaissance." *Studies in Philology* 41:1–15.

Bacon, Francis. 1955. *Selected Writings of Francis Bacon.* Ed. Hugh G. Dick. New York: Random House.

Barnes, Jonathan. 1982. *Aristotle.* Oxford: Oxford University Press.

Bennett, J. A. 1986. "The Mechanics' Philosophy and the Mechanical Philosophy." *History of Science* 24:1–28.

Biagioli, Mario. 1993. *Galileo, Courtier.* Chicago: The University of Chicago Press.

Boyle, Robert. 1965–6. *The Works of the Honourable Robert Boyle.* 6 vols. Hildesheim: Gg. Olms.

Boyle, Robert. 1992. *The Early Essays and Ethics.* Ed. John T. Harwood. Carbondale, Ill.: Southern Illinois University Press.

Brockliss, L. W. B. 1992. "The Scientific Revolution in France." In *The Scientific Revolution in National Context*, ed. Roy Porter and Mikulás Teich. Cambridge: Cambridge University Press.

Brooke, John Hedley. 1991. *Science and Religion: Some Historical Perspectives.* Cambridge: Cambridge University Press.

Brown, Harcourt. 1934. *Scientific Organizations in Seventeenth-Century France (1620–1680).* Baltimore: Williams & Wilkins.

Bruno, Giordano. 1992. *The Expulsion of the Triumphant Beast.* Trans. and ed. Arthur D. Imerti. Lincoln: University of Nebraska Press.

Burke, Peter. 1978. *Popular Culture in Early Modern Europe.* New York: Harper & Row.

Burke, Peter. 1981. *Montaigne.* Oxford: Oxford University Press.

Burtt, Edwin Arthur. 1954. *The Metaphysical Foundations of Modern Science*, 2d rev. ed. Garden City, N.Y.: Doubleday.

Bylebyl, Jerome J. 1972. "Harvey, William." In *Dictionary of Scientific Biography*, vol. 6, ed. Charles Coulston Gillispie. New York: Charles Scribner's Sons.

Caspar, Max. 1959. *Kepler*. Trans. C. Doris Hellman. London: Abelard-Schuman.

Clericuzio, Antonio. 1990. "A Redefinition of Boyle's Chemistry and Corpuscular Philosophy." *Annals of Science* 47:561–89.

Cohen, I. Bernard. 1960. *The Birth of a New Physics*. Garden City, N.Y.: Doubleday.

Collingwood, R. G. 1965. *The Idea of Nature*. Oxford: Oxford University Press.

Copenhaver, Brian P., and Charles B. Schmitt. 1992. *Renaissance Philosophy*. Oxford: Oxford University Press.

Cottingham, John et al., trans. and eds. 1991. *The Philosophical Writings of Descartes*. Volume 3: *The Correspondence*. Cambridge: Cambridge University Press.

Couliano, Ioan P. 1987. *Eros and Magic in the Renaissance*. Trans. Margaret Cook. Chicago: The University of Chicago Press.

Coward, Barry. 1992. "The Experience of the Gentry, 1640–1660." In *Town and Countryside in the English Revolution*, ed. R. C. Richardson. Manchester: Manchester University Press.

Davis, J. C. 1981. *Utopia and the Ideal Society: A Study of English Utopian Writing 1516–1700*. Cambridge: Cambridge University Press.

Dear, Peter. 1988. *Mersenne and the Learning of the Schools*. Ithaca, N.Y.: Cornell University Press.

Debus, Allen. 1975. "The Chemical Debates of the Seventeenth Century: The Reaction to Robert Fludd and Jean Baptiste van Helmont." In *Reason, Experiment, and Mysticism in the Scientific Revolution*, ed. M. L. Righini Bonelli and William R. Shea. New York: Science History Publications.

Debus, Allen G., ed. 1970. *Science and Education in the Seventeenth Century: The Webster-Ward Debate*. New York: American Elsevier.

Descartes, René. 1968. *Discourse on Method and the Meditations*. Trans. F. E. Sutcliffe. Harmondsworth, U.K.: Penguin Books.

Descartes, René. 1989. *The Passions of the Soul*. Trans. Stephen Voss. Indianapolis: Hackett.

Dobbs, Betty Jo Teeter, and Margaret C. Jacob. 1995. *Newton and the Culture of Newtonianism*. Atlantic Highlands, N.J.: Humanities Press.

Drake, Stillman, trans. and ed. 1957. *Discoveries and Opinions of Galileo*. Garden City, N.Y.: Doubleday.

Eamon, William. 1991. "Court, Academy, and Printing House: Patronage and Scientific Careers in Late-Renaissance Italy." In *Patronage and Institutions: Science, Technology, and Medicine at the European Court, 1500–1750*, ed. Bruce T. Moran. Rochester, N.Y.: Boydell Press.

Evans, R. J. W. 1973. *Rudolf II and His World*. Oxford: Clarendon Press.

Farrington, Benjamin. 1964. *The Philosophy of Francis Bacon*. Liverpool: Liverpool University Press.

Febvre, Lucien, and Henri-Jean Martin. 1984. *The Coming of the Book: The Impact of Printing 1450–1800*. London: Verso Editions.

Gascoigne, John. 1990. "A Reappraisal of the Role of Universities in the Scientific Revolution." In *Reappraisals of the Scientific Revolution*, ed. David C. Lindberg and Robert S. Westman. Cambridge: Cambridge University Press.

Gaukroger, Stephen. 1995. *Descartes: An Intellectual Biography*. Oxford: Clarendon Press.

Grant, Edward. 1981. *Much Ado about Nothing: Theories of Space and Vacuum from the Middle Ages to the Scientific Revolution.* Cambridge: Cambridge University Press.

Gregory, Tullio. 1992. "Pierre Charron's 'Scandalous Book.'" In *Atheism from the Reformation to the Enlightenment,* ed. Michael Hunter and David Wootton. Oxford: Clarendon Press.

Hale, J. R. 1994. *The Civilization of Europe in the Renaissance.* New York: Atheneum.

Hall, A. Rupert. 1981. *From Galileo to Newton.* New York: Dover.

[Hall], Marie Boas. 1962. *The Scientific Renaissance, 1450–1630.* New York: Harper & Row.

Harth, Erica. 1992. *Cartesian Women.* Ithaca, N.Y.: Cornell University Press.

Henry, John. 1986. "Occult Qualities and the Experimental Philosophy: Active Principles in Pre-Newtonian Matter Theory." *History of Science* 24:335–81.

Henry, John. 1991. "Doctors and Healers: Popular Culture and the Medical Profession." In *Science, Culture and Popular Belief in Renaissance Europe,* ed. Stephen Pumfrey et al. Manchester: Manchester University Press.

Hill, Christopher. 1972. *The World Turned Upside Down.* New York: Viking Press.

Hine, William L. 1976. "Mersenne and Vanini." *Renaissance Quarterly* 29:52–65.

Hine, William L. 1984. "Marin Mersenne: Renaissance Naturalism and Renaissance Magic." In *Occult and Scientific Mentalities in the Renaissance,* ed. Brian Vickers. Cambridge: Cambridge University Press.

Hobbes, Thomas. 1968. *Leviathan.* Ed. C. B. Macpherson. Harmondsworth, U.K.: Penguin Books.

Holton, Gerald. 1973. "Johannes Kepler's Universe: Its Physics and Metaphysics." In *Thematic Origins of Scientific Thought: Kepler to Einstein,* idem. Cambridge, Mass.: Harvard University Press.

Hull, Charles Henry, ed. 1899. *The Economic Writings of Sir William Petty.* 2 vols. Cambridge: Cambridge University Press.

Hunter, Michael. 1981. *Science and Society in Restoration England.* Cambridge: Cambridge University Press.

Hunter, Michael, and Simon Schaffer, eds. 1989. *Robert Hooke: New Studies.* Woodbridge, Suffolk, U.K.: Boydell Press.

Irwin, Terence. 1989. *Classical Thought.* Oxford: Oxford University Press.

Jacob, James R. 1978. *Robert Boyle and the English Revolution.* New York: Burt Franklin.

Jacob, James R. 1992. "The Political Economy of Science in Seventeenth-Century England." *Social Research* 59:505–32. Reprinted in Margaret C. Jacob, ed. 1994. *The Politics of Western Science 1640–1990.* Atlantic Highlands, N.J.: Humanities Press.

Jacob, Margaret C. 1991. *The Newtonians and the English Revolution, 1689–1720.* New York: Gordon and Breach.

Johnston, David. 1986. *The Rhetoric of "Leviathan": Thomas Hobbes and the Politics of Cultural Transformation.* Princeton: Princeton University Press.

Joy, Lynn Sumida. 1987. *Gassendi the Atomist.* Cambridge: Cambridge University Press.

Kaplan, Barbara Beigun. 1993. *"Divulging of Useful Truths in Physick": The Medical Agenda of Robert Boyle.* Baltimore: The Johns Hopkins University Press.

Kaufmann, Thomas DaCosta. 1993. *The Mastery of Nature: Aspects of Art, Science, and Humanism in the Renaissance.* Princeton: Princeton University Press.
Keohane, Nannerl O. 1980. *Philosophy and the State in France: The Renaissance to the Enlightenment.* Princeton: Princeton University Press.
Koyré, Alexandre. 1968. *From the Closed World to the Infinite Universe.* Baltimore: The Johns Hopkins University Press.
Kuhn, Thomas S. 1952. "Robert Boyle and Structural Chemistry in the Seventeenth Century." *Isis* 43:12–36.
Kuhn, Thomas S. 1957. *The Copernican Revolution.* New York: Random House.
Langford, Jerome J. 1971. *Galileo, Science and the Church.* Ann Arbor: University of Michigan Press.
Lapidge, Michael. 1988. "The Stoic Inheritance." In *A History of Twelfth-Century Philosophy,* ed. Peter Dronke. Cambridge: Cambridge University Press.
Lenoble, Robert. 1943. *Mersenne ou la Naissance du Mécanisme.* Paris: Vrin.
Letwin, William. 1972. "The Economic Foundations of Hobbes' Politics." In *Hobbes and Rousseau,* ed. Maurice Cranston and Richard S. Peters. Garden City, N.Y.: Doubleday.
Levi, Anthony. 1964. *French Moralists: The Theory of the Passions 1585 to 1649.* Oxford: Clarendon Press.
Lindberg, David C. 1992. *The Beginnings of Western Science.* Chicago: The University of Chicago Press.
Lloyd, G. E. R. 1970. *Early Greek Science: Thales to Aristotle.* New York: Norton.
Lloyd, G. E. R. 1979. *Magic, Reason and Experience: Studies in the Origins and Development of Greek Science.* Cambridge: Cambridge University Press.
Long, A. A. 1974. *Hellenistic Philosophy: Stoics, Epicureans, Sceptics,* 2d ed. London: Duckworth.
Lovejoy, Arthur O., and George Boas. 1980. *Primitivism and Related Ideas in Antiquity.* New York: Octagon Books.
Low, Anthony. 1985. *The Georgic Revolution.* Princeton: Princeton University Press.
Mandelbaum, Maurice. 1964. *Philosophy, Science and Sense Perception.* Baltimore: The Johns Hopkins University Press.
Martin, Julian. 1992. *Francis Bacon, the State, and the Reform of Natural Philosophy.* Cambridge: Cambridge University Press.
Merton, Robert K. 1970. *Science, Technology and Society in Seventeenth-Century England,* 2d ed. New York: Harper & Row.
Montaigne, Michel de. 1993a. *An Apology for Raymond Sebond.* Trans. M. A. Screech. Harmondsworth, U.K.: Penguin Books.
Montaigne, Michel de. 1993b. *The Essays: A Selection.* Trans. M. A. Screech. Harmondsworth, U.K.: Penguin Books.
Morgan, Vance G. 1994. *Foundations of Cartesian Ethics.* Atlantic Highlands, N.J.: Humanities Press.
Pagel, Walter. 1958. *Paracelsus: An Introduction to Philosophical Medicine in the Era of the Renaissance.* New York: S. Karger.
Paracelsus. 1951. *Selected Writings.* Ed. Jolande Jacobi. Princeton: Princeton University Press.
[Pet]t, [Pete]r. 1661. *A Discourse Concerning Liberty of Conscience.* London: For Nathaniel Brook.

Pico della Mirandola, Giovanni. 1956. *Oration on the Dignity of Man*. Trans. A. Robert Caponigri. Chicago: The University of Chicago Press.

Plattes, Gabriel. 1641. *A Description of the Famous Kingdome of Macaria*. London: Printed for Francis Constable.

Popkin, Richard H. 1979. *The History of Scepticism from Erasmus to Spinoza*. Berkeley and Los Angeles: University of California Press.

Rattansi, P. M. 1963. "Paracelsus and the Puritan Revolution." *Ambix* 11:24–32.

Redondi, Pietro. 1987. *Galileo Heretic*. Princeton: Princeton University Press.

Rivers, Isabel. 1991. *Reason, Grace, and Sentiment*. Vol. 1: *Whichcote to Wesley*. Cambridge: Cambridge University Press.

Rossi, Paolo. 1968. *Francis Bacon: From Magic to Science*. Chicago: The University of Chicago Press.

Rossi, Paolo. 1970. *Philosophy, Technology and the Arts in the Early Modern Era*. New York: Harper & Row.

Sailor, Danton B. 1964. "Moses and Atomism." *Journal of the History of Ideas* 25: 3–16.

Sarasohn, Lisa T. 1982. "The Ethical and Political Philosophy of Pierre Gassendi." *Journal of the History of Philosophy* 20:239–60.

Sarasohn, Lisa T. 1985. "Motion and Morality: Pierre Gassendi, Thomas Hobbes and the Mechanical World-View." *Journal of the History of Ideas* 46:363–79.

Sargent, Rose-Mary. 1995. *The Diffident Naturalist: Robert Boyle and the Philosophy of Experiment*. Chicago: The University of Chicago Press.

Schmitt, Charles B., and Quentin Skinner, eds. 1988. *The Cambridge History of Renaissance Philosophy*. Cambridge: Cambridge University Press.

Schneewind, J. B., ed. 1990. *Moral Philosophy from Montaigne to Kant: An Anthology*. 2 vols. Cambridge: Cambridge University Press.

Screech, M. A. 1991. *Montaigne and Melancholy: The Wisdom of the "Essays."* Harmondsworth, U.K.: Penguin Books.

Shanahan, Timothy. 1988. "God and Nature in the Thought of Robert Boyle." *Journal of the History of Philosophy* 26:547–69.

Shapin, Steven. 1988. "The House of Experiment in Seventeenth-Century England." *Isis* 79:373–404.

Shapin, Steven, and Simon Schaffer. 1985. *Leviathan and the Air-Pump: Hobbes, Boyle and the Experimental Life*. Princeton: Princeton University Press.

Shapiro, Barbara J. 1968. "Latitudinarianism and Science in Seventeenth-Century England." *Past & Present* 40:16–41. Reprinted in Charles Webster, ed. 1974. *The Intellectual Revolution of the Seventeenth Century*. London: Routledge & Kegan Paul.

Shea, William R. 1986. "Galileo and the Church." In *God and Nature*, ed. David Lindberg and Ronald Numbers. Berkeley and Los Angeles: University of California Press.

Solomon, Howard M. 1972. *Public Welfare, Science, and Propaganda in Seventeenth Century France: The Innovations of Théophraste Renaudot*. Princeton: Princeton University Press.

Spink, J. S. 1960. *French Free-Thought from Gassendi to Voltaire*. London: Athlone Press.

Stephenson, Bruce. 1994. *Kepler's Physical Astronomy*. Princeton: Princeton University Press.

Taub, Liba Chaia. 1993. *Ptolemy's Universe: The Natural Philosophical and Ethical Foundations of Ptolemy's Astronomy.* Chicago and LaSalle, Ill.: Open Court.

Thirsk, Joan. 1992. "Agrarian Problems and the English Revolution." In *Town and Countryside in the English Revolution,* ed. R. C. Richardson. Manchester: Manchester University Press.

Thoren, Victor E. 1990. *The Lord of Uraniborg: A Biography of Tycho Brahe.* Cambridge: Cambridge University Press.

Todd, Margo. 1987. *Christian Humanism and the Puritan Social Order.* Cambridge: Cambridge University Press.

Trevor-Roper, Hugh. 1967. "Three Foreigners: The Philosophers of the Puritan Revolution." In *Religion, the Reformation and Social Change,* idem. London: Macmillan.

Trevor-Roper, Hugh. 1985. "The Paracelsian Movement." In *Renaissance Essays,* idem. Chicago: The University of Chicago Press.

Tuck, Richard. 1979. *Natural Rights Theories.* Cambridge: Cambridge University Press.

Tuck, Richard. 1993. *Philosophy and Government, 1572–1651.* Cambridge: Cambridge University Press.

Tuveson, Ernest Lee. 1949. *Millennium and Utopia: A Study in the Background of the Idea of Progress.* Berkeley and Los Angeles: University of California Press.

Van Helden, Albert. 1977. *The Invention of the Telescope.* Transactions of the American Philosophical Society 67(4). Philadelphia: American Philosophical Society.

Van Helden, Albert. 1983. "The Birth of the Modern Scientific Instrument 1550–1700." In *The Uses of Science in the Age of Newton,* ed. John G. Burke. Berkeley and Los Angeles: University of California Press.

Virgil. 1982. *The Georgics.* Trans. L. P. Wilkinson. Harmondsworth, U.K.: Penguin Books.

Wear, Andrew. 1990. "The Heart and Blood from Vesalius to Harvey." In *Companion to the History of Modern Science,* ed. R. C. Olby et al. London: Routledge.

Webster, Charles. 1975. *The Great Instauration: Science, Medicine and Reform 1626–1660.* London: Duckworth.

Webster, Charles. 1979. *Utopian Planning and the Puritan Revolution: Gabriel Plattes, Samuel Hartlib and "Macaria."* Oxford: Wellcome Unit for the History of Medicine.

Webster, Charles. 1982. *From Paracelsus to Newton: Magic and the Making of Modern Science.* Cambridge: Cambridge University Press.

Webster, Charles. 1993. "Paracelsus: Medicine as Popular Protest." In *Medicine and the Reformation,* ed. Ole Peter Grell and Andrew Cunningham. London: Routledge.

Westfall, Richard S. 1977. *The Construction of Modern Science: Mechanisms and Mechanics.* Cambridge: Cambridge University Press.

Westfall, Richard S. 1980. *Never at Rest: A Biography of Isaac Newton.* Cambridge: Cambridge University Press.

Westman, Robert S. 1980. "The Astronomer's Role in the Sixteenth Century: A Preliminary Study." *History of Science* 18:105–47.

White, Andrew Dickson. 1896. *A History of the Warfare of Science with Theology in Christendom.* 2 vols. London and New York: Macmillan.

Wightman, W. P. D. 1972. *Science in a Renaissance Society*. London: Hutchinson.
Williams, George Huntston. 1992. *The Radical Reformation*, 3rd rev. ed. Kirksville, Mo.: Sixteenth Century Journal Publishers.
Yates, Frances A. 1964. *Giordano Bruno and the Hermetic Tradition*. London: Routledge and Kegan Paul.
Yates, Frances A. 1972. *The Rosicrucian Enlightenment*. London: Routledge and Kegan Paul.
Yates, Frances A. 1982. "Giordano Bruno's Religious Policy." In *Lull and Bruno. Collected Essays*, vol. 1, idem. London: Routledge and Kegan Paul.
Yates, Frances A. 1983. "The Italian Academies." In *Renaissance and Reform: The Italian Contribution. Collected Essays*, vol. 2, idem. London: Routledge and Kegan Paul.

Index